P9-CMX-122

The

Once

and

Future

Liberal

The
Once
and
Future
Liberal

AFTER IDENTITY POLITICS

Mark Lilla

HARPER

An Imprint of HarperCollins*Publishers*

THE ONCE AND FUTURE LIBERAL. Copyright © 2017 by Mark Lilla. All rights reserved. Printed in the United States of America. No part of this book may be used or reproduced in any manner whatsoever without written permission except in the case of brief quotations embodied in critical articles and reviews. For information, address HarperCollins Publishers, 195 Broadway, New York, NY 10007.

HarperCollins books may be purchased for educational, business, or sales promotional use. For information, please email the Special Markets Department at SPsales@harpercollins.com.

FIRST EDITION

Library of Congress Cataloging-in-Publication Data has been applied for.

ISBN 978-0-06-269743-1

17 18 19 20 21   LSC   10 9 8 7 6 5 4 3

We must understand that there is a difference between being a party that cares about labor and being a labor party. There is a difference between being a party that cares about women and being the women's party. And we can and we must be a party that cares about minorities without becoming a minority party. We are citizens first.

—SENATOR EDWARD M. KENNEDY (1985)

Contents

The
Once
and
Future
Liberal

Introduction:
The Abdication

Donald J. Trump is president of the United States. And his surprise victory has finally energized American liberals and progressives. They are busy organizing what they call a "resistance" to everything he stands for. They are networking, marching, showing up at town hall meetings, and flooding the phone lines of their congressional representatives. There is already excited talk about winning back House and Senate seats in midterm elections, and the presidency three years down the road. The search for candidates has begun and no doubt some staffers are already dreaming of the offices they will occupy in the West Wing of the White House.

→>—<←

The Once and Future Liberal

If only American politics were so simple. Lose the flag, capture the flag. We liberals have played this game before and sometimes won. We have had Democratic presidents in four of the ten terms that followed Ronald Reagan's victory in 1980, and there were significant policy victories during the administrations of Bill Clinton and Barack Obama. But scratch below the surface of presidential elections, which seem to follow their own historical rhythm, and things turn very dark, very fast.

Clinton and Obama were elected and then reelected with messages that were long on hope and change. But they were stymied at almost every turn by self-confident Republicans in Congress, a right-leaning Supreme Court, and a steadily growing majority of state governments in Republican hands. Those presidents' electoral victories did nothing to stop or even slow the rightward drift of American public opinion. In fact, thanks in large part to the shameless and massively influential right-wing media complex, the longer they were in office the more the public held liberalism in contempt as a political doctrine. And now we face far right populist websites mixing half-truths, lies, conspiracy theories, and fabrications into a toxic brew eagerly swallowed by the credulous, the angry, and the menacing. Liberals have become America's ideological third party, lagging behind self-declared independents and conservatives, even among young voters and certain minority groups. We have been repudiated in no uncertain terms. Donald Trump the man is, frankly, not the greatest of our worries. And if we don't look beyond him there is very little hope for us.

American liberalism in the twenty-first century is in crisis: a crisis of imagination and ambition on our side, a crisis of attachment and trust on the side of the wider public. The majority of Americans have made it abundantly clear that they no longer respond to whatever larger message we have been conveying over the past decades. And even when they vote for our candidates, they are increasingly hostile to the way we speak and write (especially about them), the way we argue, the way we campaign, the way we govern. Abraham Lincoln's famous remark is timely once again:

> *Public sentiment is everything. With it, nothing can fail; against it, nothing can succeed. Whoever*

*molds public sentiment goes deeper than he who en-
acts statutes, or pronounces judicial decisions.*

The American right understands in its bones this
basic law of democratic politics, which is why it has
effectively controlled the political agenda of this country for two generations. Liberals have for just as long
refused to accept it. Like Bartleby the Scrivener, they
"prefer not to." The question is, Why? Why would
those who claim to speak for the great American *demos*
be so indifferent to stirring its feelings and gaining its
trust? This is the question I would like to explore.

I write as a frustrated American liberal. My frustration is not directed at Trump's voters, or those who
explicitly supported the rise of this populist demagogue, or those in the press who greased the wheels
of his campaign, or those craven Washingtonians
who have fallen into line behind him. Others will take
them on. My frustration has its source in an ideology
that for decades has prevented liberals from devel-
oping an ambitious vision of America and its future
that would inspire citizens of every walk of life and in

every region of the country. A vision that would orient the Democratic Party and help it win elections and occupy our political institutions over the long term, so we might effect the changes we want and America needs. Liberals bring many things to electoral contests: values, commitment, policy proposals. What they don't bring is an image of what our shared way of life might be. Ever since the election of Ronald Reagan the American right has offered one. And it is this image—not money, not false advertising, not fearmongering, not racism—that has been the ultimate source of its strength. In the contest for the American imagination, liberals have abdicated.

The Once and Future Liberal is the story of that abdication. Its argument can be briefly summarized. I suggest that American political history over the past century can be usefully divided into two "dispensations," to invoke the Christian theological term. The first, the Roosevelt Dispensation, stretched from the era of the New Deal to the era of the civil rights movement and the Great Society in the 1960s, and then exhausted itself in the 1970s. The second, the Reagan

Dispensation, began in 1980 and is now being brought to a close by an opportunistic, unprincipled populist. Each dispensation brought with it an inspiring image of America's destiny and a distinctive catechism of doctrines that set the terms of political debate. The Roosevelt Dispensation pictured an America where citizens were involved in a collective enterprise to guard one another against risk, hardship, and the denial of fundamental rights. Its watchwords were solidarity, opportunity, and public duty. The Reagan Dispensation pictured a more individualistic America where families and small communities and businesses would flourish once freed from the shackles of the state. Its watchwords were self-reliance and minimal government. The first dispensation was political, the second anti-political.

The great liberal abdication began during the Reagan years. With the end of the Roosevelt Dispensation and the rise of a unified and ambitious right, American liberals faced a serious challenge: to develop a fresh political vision of the country's shared destiny, adapted to the new realities of American society and chastened by the failures of old approaches. Liberals failed to do this. Instead they threw themselves into the movement politics of identity, losing a sense of what we share as citizens and what binds us as a nation. An image for Roosevelt liberalism and the unions that supported it was that of two hands shaking. A recurring image of identity liberalism is that of a prism refracting a single beam of light into its constituent colors, producing a rainbow. This says it all.

The politics of identity is nothing new, certainly on the American right. What was astonishing during the Reagan Dispensation was the development of a left-wing version of it that became the de facto creed of two generations of liberal politicians, professors, schoolteachers, journalists, movement activists, and officials of the Democratic Party. This was not a historical accident. For the fascination, and then obsession, with identity did not challenge the fundamental principle of Reaganism. It reinforced that principle: individualism. Identity politics on the left was at first about large classes of people—African-Americans, women—seeking to redress major historical wrongs by mobilizing and then working through our political institutions to secure their rights. But by the 1980s it had given way to a pseudo-politics of self-regard and increasingly narrow and exclusionary self-definition that is now cultivated in our colleges and universities. The main result has been to turn young people back onto themselves, rather than turning them outward toward the wider world. It has left them unprepared to think about the common good and what must be done practically to secure it—especially the hard and unglamorous task of persuading people very different from themselves to join a common effort. Every advance of liberal *identity* consciousness has marked a retreat of liberal *political* consciousness. Without which no vision of Americans' future can be imagined.

So it should come as little surprise that the term *liberalism* leaves so many Americans indifferent if not hostile today. It is considered, with some justice, as a creed professed mainly by educated urban elites cut off from the rest of the country who see the issues of the day principally through the lens of identities, and whose efforts center on the care and feeding of hypersensitive movements that dissipate rather than focus the energies of what remains of the left. Contrary to what the centrist coroners of the 2016 election will be saying, the reason the Democrats are losing ground is not that they have drifted too far to the left. Nor, as the progressives are already insisting, is it that they have drifted too far to the right, especially on economic issues. They are losing because they have retreated into caves they have carved for themselves in the side of what once was a great mountain.

There is no clearer evidence of this retreat than the homepage of the Democratic Party. At the moment I'm writing, the homepage of the Republican site prominently features a document titled "Principles for American Renewal," which is a statement of positions on eleven different broad political issues. The list begins with the Constitution ("Our Constitution should be preserved, valued and honored") and ends with immigration ("We need an immigration system that secures our borders, upholds the law, and boosts our economy"). There is no such document to be found on the Democrats' homepage. Instead, when you scroll to the bottom of it you find a list of links titled "People." And each link takes you to a page tailored to appeal to a distinct group and identity: women, Hispanics, "ethnic Americans," the LGBT community, Native

The Once and Future Liberal

Americans, African-Americans, Asian-Americans and Pacific Islanders . . . There are seventeen such groups, and seventeen separate messages. You might think that, by some mistake, you have landed on the website of the Lebanese government—not that of a party with a vision of America's future.

But perhaps the most damning charge that can be brought against identity liberalism is that it leaves those groups it professes to care about more vulnerable than they otherwise would be. There is a good reason that liberals focus extra attention on minorities, since they are the most likely to be disenfranchised. But in a democracy the only way to meaningfully defend them—and not just make empty gestures of recognition and "celebration"—is to win elections and exercise power in the long run, at every level of government. And the only way to accomplish that is to have a message that appeals to as many people as possible and pulls them together. Identity liberalism does just the opposite.

This misorientation has real-world consequences. It is one thing to secure the constitutional right to

Introduction: The Abdication

abortion at the national level. It is quite another to guarantee that spurious barriers to obtaining one are not erected at the state and local levels. The same holds for voting rights and other matters. If, for instance, we want to protect black motorists from police abuse, or gay and lesbian couples from harassment on the street, we need state attorneys general willing to prosecute such cases, and state judges willing to enforce the law. And the only way to make sure we get them is to elect liberal Democratic governors and state legislators who will make the appointments.

But we are not even in the contest. The Republicans have successfully persuaded much of the public that they are the party of Joe Sixpack and Democrats are the party of Jessica Yogamat. The result is that today certain swaths of the country are so thoroughly dominated by the radical Republican right that certain federal laws and even constitutional protections are, practically speaking, a dead letter there. If identity liberals were thinking politically, not pseudo-politically, they would concentrate on turning that around at the local level, not on organizing yet another march in Washington or preparing yet another federal court brief. The paradox of identity liberalism is that it paralyzes the capacity to think and act in a way that would actually accomplish the things it professes to want. It is mesmerized by symbols: achieving superficial diversity in organizations, retelling history to focus on marginal and often minuscule groups, concocting inoffensive euphemisms to describe social reality, protecting young ears and eyes already accustomed to slasher films from any disturbing encounter with alternative viewpoints. Identity liberalism has ceased being a political project and has morphed into an evangelical one. The difference is this: evangelism is about speaking truth to power. Politics is about seizing power to defend the truth.

There can be no liberal politics without a sense of *we*—of what we are as citizens and what we owe each other. If liberals hope ever to recapture America's imagination and become a dominant force across the country, it will not be enough to beat the Republicans at flattering the vanity of the mythical Joe Sixpack. They must offer a vision of our common destiny based on one thing that all Americans, of every background, actually share. And that is citizenship. We must re-learn how to speak to citizens *as citizens* and to frame our appeals—including ones to benefit particular groups—in terms of principles that everyone can affirm. Ours must become a civic liberalism.*

This does not mean a return to the New Deal. Future liberals cannot be like the liberals of yore; too much has changed. But it will require that the spell of identity politics that has held two generations in its thrall be broken so that we can focus on what we share as citizens. I hope to convince my fellow liberals that their current way of looking at the country, speaking to it, teaching the young, and engaging in practical politics has been misguided and counterproductive. Their abdication must end and a new approach must be embraced.

---

* It is a sign of how polluted our political discourse has become that any mention of the term *citizen* leads people to think of the hypocritical and racist demagoguery that passes for our "debate" on immigration and refugees today. I will not be discussing such matters here, and what I have to say about citizenship implies nothing about who should be granted citizenship or how noncitizens should be treated.

The Once and Future Liberal

→>‹‹

It is a bittersweet truth that there has never been a bet-ter opportunity in half a century for liberals to start winning the country back. Republicans since Trump's election are in disarray and intellectually bankrupt. Most Americans now recognize that Reagan's "shining city upon a hill" has turned into rust belt towns with long-shuttered shops, abandoned factories invaded by local grasses, cities where the water is undrink-able and guns are everywhere, and homes across the country where families are scraping by with part-time minimum-wage jobs and no health insurance. It is an America where Democrats, independents, and many Republican voters feel themselves abandoned by their country. They want America to be America again.

But there is no *again* in politics, just the future. And there is no reason why the American future should not be a liberal one. Our message can and should be sim-ple: we are a republic, not a campsite. Citizens are not roadkill. They are not collateral damage. They are not the tail of the distribution. A citizen, simply by virtue of being a citizen, is one of us. We have stood together to defend the country against foreign adversaries in the past. Now we must stand together at home to make sure that none of us faces the risk of being left behind. We're all Americans and we owe that to each other. That's what liberalism means.

American liberals have a reputation, as the saying goes, of never missing an opportunity to miss an opportunity. May that prophecy not be fulfilled this time. The election of Donald Trump has released stores of energy among liberals and progressives that even they seem surprised to have discovered within themselves. A popular wave from the left has risen up to resist a populist one from the right, and it's encouraging to observe. But "resistance" will not be enough. Our short-term strategy must be to direct every bit of that energy into electoral politics so we can actually bring about the change we profess to seek. And our longer-term ambition must be to develop a vision of America that emerges authentically out of liberal values yet speaks to every citizen, as a citizen. This will require a reorientation of our thinking and engagement, but above all it will mean putting the age of identity behind us. It is time—past time—to get real.

I

Anti-Politics

‎‎

I see an immense crowd of similar and equal men who spin restlessly around themselves, seeking vulgar little pleasures to fill their souls. Living apart, each is like a foreigner to the fate of others. His children and friends are for him the entire human race. As for his fellow citizens, he is next to them but does not see them, he touches them but does not feel them. He exists only in and for himself, alone. And though he may still have a family, he no longer has a country.

—Alexis de Tocqueville

My ideal citizen is the self-employed, homeschooling, IRA-owning guy with a concealed-carry permit. Because that person doesn't need the goddamn government for anything.

—Grover Norquist

A Word from Karl Marx

I n January 1981 a new president leading a rejuvenated party was sworn into office. It felt, though, as if something more significant than an election had just taken place. *Old things are passed away; behold, all things are become new.* It is difficult to convey to anyone who wasn't alive and politically aware at the time what a dreary place America seemed in the late 1970s, how lacking in direction and confidence. All the slogans inherited from the New Deal and Great Society, all the old convictions, all the old approaches, bore a deathly pallor. They convinced and motivated no one. A new generation of well-educated conservative intellectuals had serious, fresh ideas about reforming (not abolishing) government that they really believed in, making them seem the brightest kids in the class. But it wasn't careful study of their arguments that convinced millions of Americans to vote for Ronald Reagan. It was the imaginative connection he made with the public that transformed those ideas into an epiphany, a vision of a new way of national life, masquerading as an old one. And this allowed him to cast himself as a homespun, *aw-shucks* John the Baptist. In the year before the election seven out of eight Americans said they were dissatisfied with the way things were going in the country. By 1986 only a quarter felt that way. Halleluiah.

Liberal Democrats scoffed—their first mistake. They just couldn't wrap their minds around the fact that with Reagan they were not up against retread versions of Cold War and Rotary Club pieties, recited by a puppet of the pin-striped classes. They were up against a new political dispensation. A political dispensation is a difficult thing to define, and difficult even to perceive until it draws to an end and the gap between rhetoric and reality becomes suddenly apparent. A dispensation is not grounded in a set of principles or arguments; it is grounded in feelings and perceptions that give principles and arguments psychological force. When the Great Depression struck, sensible talk about fiscal responsibility and balanced budgets trailed off and discussions about the fate of the "forgotten man" became more intense. The budget was still there, just as the forgotten man had been there throughout the Gilded Age. Yet somehow everything looked and felt different. With the election of FDR a switch was flipped, a tipping point reached, a gestalt shifted—choose your metaphor. And once that happens there is no going back. If you are unhappy with the terms of debate during one dispensation, you have no choice but to prepare a new one. Nostalgia is suicidal.

So is the lazy thought that you just have to wait until people rise up against the usurpers. Republicans made that mistake with Roosevelt, and Democrats made it with Reagan. We must never forget that moving hearts and minds for more than one election cycle is not easy, and if an ideology endures, this means that it is capturing something important in social reality. Marx was right about this:

material conditions help to determine which political ideas resonate at any historical moment. There were concrete material reasons why the Roosevelt Dispensation lasted for four decades, and new concrete material reasons why the Reagan Dispensation lasted about as long. Which means that if liberals are serious about supplanting Reaganism in the public imagination, they must first understand why it arose and retained its power to convince for so long. What changes in American economic and social life made that ideology plausible in the first place?

Answering this question is an important exercise. Whatever vision of America and its future liberals eventually offer, it must be based on a coldly realistic view of how we live now. We go into politics with the country we have, not the country we might wish for. Reaganism endured because it did not declare war on the way most Americans were living and thinking about themselves. It fitted right in. And it has lost force because the contradiction between the dogmas and social reality is becoming all too apparent. The same is true of identity liberalism. It gained force because it was in harmony with some of the deep social changes that Reaganism responded to, as well. And now, given where the country is, we need something else. But first we need to understand how we got here.

Elementary Particles

One revolution can obscure another. The year 1989 stands out in historical memory as the moment when the Soviet empire collapsed, and with it all the hopes invested in communism and revolutionary politics around the world. To some American officials and commentators it even seemed that liberal democracy had been crowned as the last political ideology standing, and perhaps even as the unconscious goal of all human striving in history. It was a triumphalist period, and therefore full of ironies.

The largest was that as the idea of democratic politics was advancing around the world in the last decade of the Cold War, Americans were becoming less and less invested in the practice of it. After the walls fell in Eastern Europe, the serious business of constitution writing began and there were earnest debates about everything from the relative powers of the executive, legislative, and judicial branches to which basic rights and social guarantees should be enshrined into law. New parties formed, then factions within them, which then split off and became even newer parties. All this was an extraordinary experience for peoples who had been prevented from determining their collective destinies for generations. They were finally citizens.

In the United States the picture was very different. Though Ronald Reagan publicly supported pro-democracy groups like Solidarity in Poland and dramatically called on Soviet leader Mikhail Gorbachev to tear down the Berlin Wall, at home he had been elected by people who could no longer quite see the point of arguing about the common good and engaging politically to achieve it. A new outlook on life had been gaining ground in the United States, one in which the needs and desires of individuals were given near-absolute priority over those of society. This subliminal revolution has done more to shape American politics in the past half century than any particular historical event.

→►◄←

Anti-Politics

Every revolution has material preconditions, and this one did too. The thirty years of uninterrupted economic growth and technological advance that followed the Second World War had no historical precedent. Rising wages and public policies encouraging home and car ownership triggered the vast growth of suburbs surrounding major American cities and then a slow shift of population to the South and West. As people left the old familiar neighborhoods of families and friends—and contact with the social problems that beset any big city—they found themselves in what seemed like virgin territory, surrounded by others to whom they had no connection and who seemed to be only passing through, like themselves. It was a toy frontier with comfortable settlers living in split-level, air-conditioned wagons. Virtually every aspect of middle-class life was transformed out there.

Consider the family. Thanks to new home appliances and the automobile, homemakers in the 1950s found themselves more independent and freed from some drudgery, but also more isolated and far from job opportunities. By the 1960s the frustrated housewife became a stock character in our literature and movies, and soon *la pasionaria* of a new wave of feminism. The birth control pill, no-fault divorce, and legalized abortion gave husbands and wives erotic independence from each other. Unsurprisingly, divorce rates spiked, and men and women got married later, or not at all. A growing number of mothers, also unsurprisingly, soon found themselves struggling to raise children by themselves. Over subsequent decades life changed for the children too. They had fewer siblings, so they got used to spending a lot of time alone or being ferried around in SUVs outfitted to withstand Operation Desert Storm. They lived in actual or de facto gated communities where they never learned to take a stroll and meet people, and where the sight of a lone child walking to school would bring concerned calls to the police, who would scold the parents for taking such a risk. Eventually the kids went to college, often far away, and after graduation joined the new urban class of independent twenty- and thirtysomethings with no responsibilities to anyone but themselves. They visited their parents and siblings on rushed visits over the holidays; otherwise they just kept in contact online. Until they finally got married, moved to the suburbs, and the whole cycle began again.

→►◄◄

We have become a hyperindividualistic bourgeois society, materially and in our cultural dogmas. Almost all the ideas or beliefs or feelings that once muted the perennial American demand for individual autonomy have evaporated. *Personal* choice. *Individual* rights. *Self*-definition. We speak these words as if a wedding vow. We hear them in school, we hear them on television, we hear them in stuffy Wall Street boardrooms, we hear them in light-filled Silicon Valley playpens, we hear them in church—we even hear them in bed. We hear them so often that it's hard for us to think or talk about any subject except in these self-regarding terms. And so it was to be expected that eventually our politics would catch up and be infected with this same self-regard, and that our political vocabulary would be revised to match the new reality. In 1974 the Harvard philosopher Robert Nozick published a bestselling book titled *Anarchy, State, and Utopia*. He shocked many by arguing that if we take the principle of individual rights seriously, then only a very minimal state could possibly be justified. What did not shock them was the argument's unquestioned assumption, which they shared:

There is no social entity with a good that undergoes some sacrifice for its own good. There are only individual people, different individual people, with their own individual lives. Using one of these people for the benefit of others, uses him and benefits the others. Nothing more. . . . Talk of an overall social good covers this up.

The very concept of *we* now seemed suspect. It is hard to imagine Ronald Reagan having read Robert Nozick, but he didn't have to. They were breathing the same cultural air, so to speak. But it was the actor not the philosopher who made the leap from potentiality to actuality and transformed an austere libertarian creed into an imaginative vision of the good life that could be had if only he was elected. And he was.

Every political dispensation comes with a catechism, and Reagan's was not difficult to memorize. It had four simple articles of faith:

- That the good life is that of self-reliant individuals—individuals embedded perhaps in families, churches, and small communities, but not citizens of a republic with common goals and duties to each other.

- That priority must be given to building rather than redistributing wealth, which allows individuals and families to maintain their independence and flourish.

- That the freer markets are, the more they will grow and enrich everyone.

- That government, to quote Reagan, "is the problem." Not tyrannical government, or inefficient government, or unjust government. Government itself.

This catechism is not conservative in any traditional sense. It treats as axiomatic the primacy of self-determination over traditional ties of dependence and obligation. It has next to nothing to say about the natural needs of collectivities—from families to nations—or our obligation to meet them. It has a vocabulary for discussing mine and thine, but not for invoking the common good or addressing class or other social realities. Its picture of our existence is that of elementary particles spread out in space, each rotating at its own speed and following its own trajectory.[*]

Sunrise

A ll this was a profound break from the catechism of the Roosevelt Dispensation. Many liberals think

[*] A social conservative might object to this assertion on the grounds that I have left out the moral education provided by the churches. What is striking about American religion during the Reagan Dispensation, though, is the degree to which faith was adapted to the ambient libertarianism rather than softening it. Before suburbanization, mainline Christian churches had thrived in ethnic urban neighborhoods and rural small towns where people knew each other. In the suburbs people began drifting away or they joined new evangelical groups whose doctrines were remarkably free of dogma, guilt, and social obligation. Over time, committing to even one of these churches proved too constraining and Americans got into the habit of "grazing," attending different churches on different Sundays depending on how the mood struck them. More and more got saved, but alone.

of the change as a kind of moral fall into selfishness and so imagine that if only Americans became "better people" they would return to the Democratic family. What this fantasy ignores is that the two dispensations arose out of very different social realities and historical experiences. The Roosevelt Dispensation was first embraced in the 1930s because it responded to the evident failure of conservatives to employ government to confront the two great challenges of the day: economic collapse and the spread of fascism. Under FDR's leadership, the experience of danger faced and overcome in the Great Depression and the Second World War bound the country together in a way it had never been bound before. It was this new social fact, not moral conversion, that allowed liberals to develop an inspiring catechism that was professed, or simply assumed, by most Americans for nearly half a century.*

Behind that catechism was a political vision of what the country was and what it might become. It was class based, though it included in the deserving

---

* Including many Republicans. Recall that Richard Nixon created a vast network of federal grants to state and local governments for social programs, set up an imposing agency to regulate

class people of any walk of life—farmers, factory workers, widows and their children, Protestants and Catholics, Northerners and Southerners—who suffered from the scourges of the day. In short, nearly everyone (though African-Americans were effectively disenfranchised in many programs due to Dixiecrat resistance). The Roosevelt Dispensation also coincided with the spread of mass media—the glossy weeklies, newsreels, movies, later television—which for fifty years projected far and wide images that corresponded to the Roosevelt vision. Arresting pictures of injustice got embedded in the American mind: caravans of dust bowl farmers heading west, dilapidated shacks in Appalachia, striking workers advancing against police charges, African-Americans being bitten by dogs and hosed down for daring to order food at a lunch counter.

air and water emissions, and another to regulate workers' health and safety. He also tried to establish a guaranteed minimum income for all working families, and to top it off proposed a national health plan that would have provided government insurance for low-income families, required employers to cover all their workers, and set standards for private insurance.

But there were also images of Americans working together to improve the country, and even the world: soldiers raising a flag after defeating a fascist army, their wives in overalls running lathes back home, shirtless workers building Hoover Dam, power and phone cables strung over the mountains, returning veterans on college campuses, citizens locking arms to demand voting rights, and young Peace Corps volunteers spreading American goodwill abroad. In Roosevelt's vision, four universal freedoms were declared and accepted as obvious by most people: freedom of speech, freedom of worship, freedom from want, and freedom from fear. This vision filled three generations of liberals with confidence, hope, pride, and a spirit of self-sacrifice. And patriotism. They had no problem standing for the national anthem.

But every catechism tends over time to become rigid and formulaic, until it eventually becomes detached from social reality. Which is exactly what happened to American liberalism in the 1970s. To the principle that collective action to serve the public good was legitimate, it added the profession of faith that taxes, spending, regulations, and court decisions were always the best way to accomplish this. By the 1980s there were countless reasons to question the assumption that government knew what it was doing and could be trusted to do it—Vietnam, Watergate, impotence in the face of stagflation, and more. Too many programs were introduced in the Great Society, too quickly and with rhetoric so elevated that it created exaggerated expectations, which resulted in inevitable disappointment. Frustratingly, none of these programs seemed capable of reversing the decline of big cities and the expansion of the welfare rolls. And some programs clearly made matters worse. Compounding the problem was that liberals refused to speak about the new culture of dependency, or about the tremendous rise in violent crime in the 1960s, most of it having nothing to do with drug offenses. Instead of seeing what they saw, they went after those who were allegedly "blaming the victim," the title of a highly influential liberal book of the 1970s, and so lost credibility with white lower-middle-class voters deemed guilty of this meta-crime.

Well-meant regulations imposed without coordination by dozens of agencies were engulfing small businesses and beginning to stifle economic growth. Liberals stood unquestioningly behind unions, even when they resisted reasonable adaptations to new technology or just defended privileges nonunionized workers could only dream of. Most foolishly, liberals grew increasingly reliant on the courts to circumvent the legislative process when it failed to deliver what they wanted (and I wanted too). Decisions rained down on everything from protecting rare fish to more explosive matters, such as abortion and school busing. Liberals lost the habit of taking the temperature of public opinion, building consensus, and taking small steps. This made the public more and more susceptible to the right's claim that the judiciary was just an imperial preserve of educated elites. The charge stuck and the approval of judicial nominations has ever since been a highly partisan process, which the right now dominates. All these factors combined to convince a growing number of Americans that even if they wanted to work together, government action would be ineffective, too costly, counterproductive, or uncontrolled.

→►◄←

Enter Ronald Reagan. Roosevelt's political vision was no longer compelling to members of the relatively affluent, hyperindividualized, and suburbanized society America had become. Americans no longer felt they needed one another as much or owed much to one another. And so Reagan offered them a new, antipolitical conception of the good life that reinforced what they were experiencing out on the new frontier. He drew on old tropes already in the country's imagination from the pre-Roosevelt days: images of the self-reliant settler and yeoman farmer, of families saying grace over a meal, of simple virtue threatened by urban life, of a self-serving professional elite exploiting the less educated, and of a strong military resisting a clear and present danger. But he also subtly updated these images for a new and very different class of mainly white Americans. Not rural families drawing water from backyard pumps as the cows mooed, but residents of subdivisions where the only sound to be heard in the afternoon was the *chik, chik, chik* of sprinkler systems. People who had attended college, however briefly, and who worked in office parks or hospitals, not out on the range. The gap between the images and reality was large, but in a way that worked in Reagan's favor. His vision was simultaneously nostalgic and futuristic: it convinced Americans that the happiness of the golden age was still attainable, just over the next ridge, if the goodness and creative energies of the country could only be unleashed.

Reagan abandoned the dour, scolding, apocalyptic style of 1950s conservatism and radiated hopefulness. After George McGovern's lame plea, *Come home America!*, after Jimmy Carter's sensible shoes and sensible sweater and sensible advice to lower the thermostat, Reagan beamed. "Twilight? Not in America. Here it's sunrise every day." More important, he exuded admiration for Americans and didn't ask them to change a thing. After Jimmy Carter delivered his diagnosis of America's malaise, Reagan responded, "I find no national malaise. I find nothing wrong with the American people." He even had the daring to tell voters they should reelect Carter "if he instills in you pride for your country and a sense of optimism about our future"—a brilliant parry that just reminded people how much they wanted to feel patriotic again.

But the new patriotism was not political, and certainly not connected in any way with government. Even military service was "marketed" differently during the Reagan Dispensation. The Army recruiting slogan introduced in 1980 was *Be all you can be!*—which could have been lifted from a Dale Carnegie book. Televised recruiting commercials centered on skills training and job opportunities that might open up in the business world after active service, not on experiences of camaraderie and self-sacrifice to be had while the recruits were in uniform. During the George W. Bush administration that slogan was changed again to *An Army of One!*—which was more bellicose but no less individualistic.

The word *government* took on a weird echo during the Reagan Dispensation, as it has periodically in American history. When Republicans uttered it listeners had conjured in their minds the image of an alien spaceship descending on the happy residents of

Middlesuburb, U.S.A., sucking up into itself all the resources, corrupting the children, and enslaving the population. You would never think to listen to them that we live in a democratic system where we get to elect representatives, toss them out if we aren't happy with them, speak up at town meetings, and appeal all the way to the Supreme Court if we feel our rights have been violated. You would never guess that the phrase they love to repeat, *We, the people*, stands at the beginning of a document that establishes a new, yes, government. You would never guess that without government, their lives and those of their voters would grind to a halt. Who, after all, sends all those Social Security and Medicare checks to elderly Republicans?*

---

* Some years ago that bible of good sense *The Onion* published an item with the title "Libertarian Reluctantly Calls Fire Department":

CHEYENNE, WY—After attempting to contain a living-room blaze started by a cigarette, card-carrying Libertarian Trent Jacobs reluctantly called the Cheyenne Fire Department Monday. "Although the community would do better to rely on an efficient, free-market fire-fighting service, the fact is that expensive, unnecessary public fire departments do exist," Jacobs said. "Also, my house was burning down."

+>-<+

Reagan ignored all that. The good life, he promised America, would emerge spontaneously as individuals and families went about their private business—especially business. A new American hero was born, The Entrepreneur. The cult that arose to worship him in the 1980s offered dreams of an easy path to nobility, open to anyone with an idea, a garage, and a few power tools. Easy in another sense, too, in that it made no moral demands. Americans have always been entrepreneurial and have always believed that to get rich is glorious. But our long-abandoned Calvinism treated wealth as a sign of moral worth, the fruit of discipline and self-denial, not the fruit of looking out for number one. The Horatio Alger stories were not Gordon Gekko stories or Ivan Boesky stories or Bernie Madoff stories. The characters did wear suspenders, but they were not masters of any universe, did not smoke big cigars or drink $1,000 bottles of wine or take clients to strip clubs. For all his social conservatism, Ronald Reagan's vision of the good life was remarkably amoral. He did not explicitly preach or encourage hedonism; he did not extol the culture of impunity that developed during his presidency. But neither did he criticize it. He understood our libertarian society too well to make that mistake.

Sunset

B y evoking an image of a better, morally undemanding life in a less political America, Reagan managed to unite the Republican Party, which after Watergate was a fractious and undisciplined body, much as the Democratic Party is today. It grouped together liberal patricians from the East, resentful Southerners and Midwestern blue-collar ethnics who had abandoned the Democratic Party, single-minded free marketeers, anti-communist crusaders, unhinged conspiracy theorists, religious leaders repelled by the cultural changes of the 1960s, and—a not insignificant group—conservative women who considered feminism an attack on themselves as mothers and homemakers. It was an ideologically and temperamentally diverse coalition. But it lacked a common vision of what America was and could become. When Reagan provided one, the party ceased to be a coalition and became an ideologically unified and electorally potent force that thought and acted like a "fine-tuned machine," to borrow a characteristic phrase from our current president. And until recently it has been.

Once Reagan was elected, the Republican strategy had two components. The first was to build from the bottom up, getting the party rooted so it could win state and local elections, then congressional elections, then the presidency. When it comes to the presidency, liberal Democrats have daddy issues, even when their candidate is a woman. Rather than concentrate on the daily task of winning over people at the local level, they have concentrated on the national media and invested their energies in trying to win the presidency every four years. And once they do, they expect Daddy to solve all the country's problems, oblivious to the fact that without support in Congress and the states a president under our system can accomplish very little. And so they are perpetually dissatisfied with their presidents and snipe at them from the left, which is the last thing a Democratic president in the current environment needs. Republican thinking from the start has been very different, as Grover Norquist, the influential president of Americans for Tax Reform, openly and amusingly admits:

> *We are not auditioning for fearless leader. We don't need a president to tell us in what direction to go. We know what direction to go. . . . We just need a president to sign this stuff. We don't need someone to think it up or design it. The leadership now for the modern conservative movement for the next twenty years will be coming out of the House and the Senate. . . . Pick a Republican with enough working digits to handle a pen to become president of the United States.*

Mission accomplished.

The other component was to build cadres through political education. Republicans sought out wealthy donors to set up foundations and think tanks as safe spaces outside the university for elaborating the Reagan catechism, a document that grew from a cocktail napkin to a vast library of popular books and academic policy studies. They set up summer camps where college students could read Aristotle and Alexander Hamilton and Friedrich von Hayek, and learn to connect them. They set up reading groups for professors, who got paid to attend. They funded graduate students and apprenticed them under movement-approved professors. They also funded campus newspapers and national organizations like the Federalist Society, which introduces students to the "originalist" interpretation of constitutional law and acts as an employment agency for young lawyers looking for clerkships and teaching positions. This one organization has revolutionized the way law is taught and interpreted in this country, and therefore how we are governed. It is the fruit of the conservatives' pedagogical strategy. The movement's fathers and godfathers, some of whom had once been Trotskyites, understood intuitively that to make lasting change the movement would have to build and sustain cadres, and send them out with full backpacks on the long march through the institutions. Marching with the aim of dismantling government by first seizing control of it, thus achieving anti-political ends by political means.

→>-<+-

In the 1980s the Reagan right reaped what it had carefully sown. But, as happened to liberals in the 1970s, proselytes soon began to radicalize the catechism and overreach, until the dogmas started becoming detached from reality. The radicalization of Reaganism happened very quickly, and very strangely once the genial Reagan was gone. The moment Bill Clinton was elected president, the conservative movement fell into a fit of hysteria reminiscent of the 1950s and which has yet to pass. It didn't matter that Clinton had fought to move the Democratic Party to the center, that with regard to economics and foreign policy he was a realist, that he had declared that "the era of big government was over" and called to "end welfare as we know it." Congressional Republicans wanted to stymie him no matter what it took. They shut down the government just because, and impeached Clinton for a peccadillo. Unlike Ronald Reagan, they became absolutists on tax cuts, gun control, and abortion, and purged from their ranks any who dared dissent. And they began participating in the daily bacchic rites of shock radio and Fox News, whipping each other and their base into a frenzy of apocalyptic doom about the state of the country. Morning in America? No, midnight! *Midnight! Midnight!*

The impending conservative crack-up seemed obvious to cooler heads in the movement, and during the Clinton years several published excellent books on what was going wrong and how to get back on course. And when George W. Bush ran on a platform of "compassionate conservatism" many welcomed what they took to be a more civic-minded vision. But within a year that rhetoric was abandoned, and not just because of the terror attacks of 9/11. The Jacobins of the conservative movement, funded by a new class of fanatical billionaires with no experience in Washington, were in control and purifying the ranks. Two-handed scholars—"on the one hand, on the other"—were purged from the think tanks, which were dropping any pretense of independence and simply pledged allegiance to the right wing of the Republican Party. The level of right-wing radio and television, such as it was, sank even further and the grunts of aggressive know-nothings drowned out real conversation. The

Republican Party's gerrymandering efforts were also having unintended consequences. For years the party had poured resources into state legislatures in order to redraw congressional districts and lock in Republican control. But this strategy also left incumbents susceptible to primary challenges by candidates who were more radical than they themselves were and who had their pick of billionaires to fund their campaigns.

And yet the machine kept churning. The Republican establishment remained calm because it was convinced it had an unbeatable strategy. If the policies it promoted and instituted were successful, all was well. If they failed, Republicans could always claim that the Washington deep state and the media were still in enemy hands, so more extreme measures were called for. This is a classic ploy familiar to revolutionary leaders throughout history: the failure of the revolution proves the need to radicalize it. This is why, for decades now, Americans have been spectators at the dark comedy of Republicans running for office successfully against "the government"— and then, once in power, running for reelection on the promise to bring down "the government" they themselves control.

The Once and Future Liberal

→>-<+

Then, as the Bush years drew to a close, the Reaganite Republican establishment found itself confronted by two major challenges. The first was Barack Obama, who had just trounced John McCain. Obama was a new face in more than the obvious ways. He was not perceived to be a triangulator as Bill Clinton was (on the far left as well as on the far right). He praised the force of Reagan's hopeful rhetoric and tried to fashion his own, though with only moderate success. Admittedly, it was a little hollow. (*Hope* . . . in what? *Yes we can!* . . . do what?) And it failed to conjure up memorable images of America's past and future, as FDR and Reagan always could. But to millions of Americans, especially young ones who grew up in the inarticulate and war-torn Bush years, it sounded like poetry. The grammatically challenged rants of Sarah Palin did not, nor did the rude cheers of Joe the Plumber.

The other challenge came with the surge of populist anger following the 2008 recession. Some of that anger was channeled into the Tea Party, which was quickly absorbed into the Republican army, where it provided shock troops in counteroffensives against every one of Obama's initiatives and appointments. But it was really the then-popular right-wing demagogue Glenn Beck who gave Republicans a taste of what was to come as the recession deepened. Beck was an apocalyptic yet strangely ebullient conspiracy theorist who on his daily Fox News broadcasts filled blackboard after blackboard with crazy Venn diagrams exposing the hidden links between 1960s radicals and Barack Obama. But he also broke with many Republican dogmas, particularly on economics and foreign policy, writing in one of his books, "Under President Bush, politics and global corporations dictated much of our economic and border policy. Nation building and internationalism also played a huge role in our move away from the founding principles." Beck's economic nationalism and isolationism struck a chord with the public, and many flocked to his sold-out rallies to hear him denounce phantom leftists but also Wall Street and the big banks. He even wrote a bestselling thriller in which all these evil forces join hands to squelch American liberty. For all his bombast, Beck was among the first on the right to report the truth that the American middle class was being hollowed out and that its children faced drastically reduced prospects. That a small class of highly educated people was benefiting from the new global economy and becoming fantastically wealthy. And that vast sections of the country had become deserted, heartbroken . . . and angry. Mainstream Republicans never got the message. Donald Trump did.

The Republican primaries of 2016 will no doubt prove as historically significant as the election that followed. We must never forget that Trump defeated both of America's major political parties, starting with the one he nominally belonged to. It was an extraordinary spectacle. The Breaker of the Idols did not come from the left or from the right. He came from below. He was unconstrained by piety toward the Gipper, by fealty toward the cause, by deep study of the Laffer Curve, or by adherence to the principle of noncontradiction. He spoke truth more often than his critics gave him credit for, but the way a child sometimes does, by accident, and then embarrasses the adults in the room. Standing before closed manufacturing plants and crowds of unemployed workers he declared that free labor markets and trade agreements were destroying more wealth than they created for such people. He spoke without hesitation about training them and providing them with some minimal health care guarantees. He spoke as if America owed it to them. (Sensing the audience's mood, though, he did not mention what they owed each other.) The other candidates just stared at their shoes. In one respect only was Donald Trump indebted to the conservative movement that Ronald Reagan founded: it bequeathed to him an angry, fearful base, which he was even more adept at manipulating than movement leaders had been. He destroyed his Republican adversaries by out-apocalypsing them.

Is this the beginning of the end of the Reagan Dispensation? There are reasons to think so. Indeed, there are reasons to think that we are already in an interregnum. Political scientists sometimes speak of

The Once and Future Liberal

"disjunctive" presidencies, which are those that mark the end of one era without inaugurating a new one. Jimmy Carter's presidency now looks in hindsight to have been disjunctive, bringing to a close the Roosevelt Dispensation and preparing the ground for Reagan's. Though Trump's election was a momentous defeat for Democrats and threatens everything liberals have ever worked for, it also exposed the emptiness of anti-political conservatism. It is hard to imagine it returning in anything like its original Reaganite form. But that is no reason for complacency. It is easy to imagine that until liberals succeed in recapturing the country's imagination, a new class of populist demagogues drawing selectively from the Reagan catechism and even radicalizing some of its dogmas will still be able to stir up and exploit public anger. They already are.

When the German tribes finally occupied ancient Rome for good in the fifth century CE, they began to practice what historians call spoliation. Unschooled in the principles of architecture and the craft of sculpture, the new Romans began ripping columns and pilasters and architraves from old temples and public edifices, and then slapped them somewhat randomly onto crude buildings of their own making, to give the structures what they took to be an imperial air. The results could be quite comical. But some are still standing.

II

Pseudo-Politics

This focusing upon our own oppression is embodied in the concept of identity politics. We believe that the most profound and potentially most radical politics come directly out of our own identity, as opposed to working to end somebody else's oppression.

—"The Combahee River Collective Statement" (1977)

Forms of ID

Thus, a short history of the Reagan Dispensation. Or half of it. The other half is about how American liberals responded to the new era they found themselves in. It is not a happy tale.

You might have thought that, faced with a novel anti-political picture of the nation, liberals would have countered with an imaginative, hopeful vision of what we share as Americans and what we might accomplish together. Instead, they lost themselves in the thickets of identity politics and developed a resentful, disuniting rhetoric of difference to match it. You might have thought that, faced with Republicans' steady acquisition of institutional power, they would have poured their energies into helping the Democratic Party win elections at every level of government and in every region of the country, reaching out especially to working-class Americans who used to vote for it. Instead, they became enthralled with social movements operating outside those institutions and developed disdain for the *demos* living between the coasts. You might have thought that, faced with the dogma of radical economic individualism that Reaganism normalized, liberals would have used their positions in our educational institutions to teach young people that they share a destiny with all their fellow citizens and have duties toward them. Instead, they trained students to be spelunkers of their personal identities and left them incurious about the world outside their heads. You might have thought a lot of reasonable things. And you would have been wrong.

There is a mystery at the core of every suicide. But a backstory can be told about all the conditions and events and choices that set the stage for the ultimate dénouement. The story of how a successful liberal politics of solidarity became a failed pseudo-politics of identity is not a simple one. It involves deep changes in American society that took place after the Second World War, the surge of political romanticism unleashed by opposition to the Vietnam War in the 1960s, the retreat of the New Left into American universities, and more.

My version of the story places special emphasis on the universities, and for a reason. Up until the 1960s, those active in liberal and progressive politics were drawn largely from the working class or farm communities, and were formed in local political clubs or on shop floors. That world is gone. Today's activists and leaders are formed almost exclusively in our colleges and universities, as are members of the mainly liberal professions of law, journalism, and education. Liberal political education now takes place, if it takes place at all, on campuses that are largely detached socially and geographically from the rest of the country—and in particular from the sorts of people who once were the foundation of the Democratic Party. This is not likely to change. Which means that liberalism's prospects will depend in no small measure on what happens in our institutions of higher education.

→>-<+-

But what exactly do we mean by identity? It is a commonplace today that identity has always played a role in American politics. If by this people mean to refer to racism, xenophobia, misogyny, and homophobia, they are correct. But interestingly, the term *identity*—in the contemporary sense of an inner thing, a homunculus that needs tending to—did not enter American political discourse until the late 1960s. It is more exact to say that the founding problem of the United States was that of political identification, beginning in colonial times.

The Pilgrims and other religious dissenters who fled England for our shores did not speak in terms of personal identities; they had souls back then. What they were seeking in America, though, was a place where they could fully identify with the country, while still fully identifying with whatever church they chose. The consensus in Europe, especially after the Wars of Religion, was that such dual identification was a psychological impossibility, given Christianity's ambiguous relation to political life. But it turned out not to be impossible in America, because the principles the country was founded on gave Christians reasons to identify with the state *because* the state guaranteed their right to identify with their churches. That proved the trick. And so, in a sense, to become an American you had to identify only with one thing: the American system of religious liberty. The citizenship bond took logical precedence because without it the Christian bond could not be protected.

A similar dynamic of dual identification has been at work in the history of immigration in the United States. The country was founded with the implicit assumption of Anglo-Protestant cultural dominance, which was threatened by steadily rising waves of immigration beginning in the nineteenth century. And just as in antiquity there had been disputes over whether Christians could be good Romans, now ones broke out over the loyalty of so-called hyphenated Americans and their commitment to act as loyal citizens—and not, say, as agents of the Pope or the German kaiser. The xenophobes, in a classic case of projection, said that ethnic loyalties would always trump democratic ones, so immigration had to be limited if not eliminated. Others argued that new arrivals could become citizens but only if their families assimilated fully to Anglo-Protestant cultural ways. Others still, like Theodore Roosevelt, thought that a "new American" type had to be forged in the melting pot, into which even Anglo-Protestants would have to jump. By the mid-twentieth century, a little assimilating and a little melting had both taken place. But it also became clear to just about everyone that neither would fully succeed—and that this was a good thing, too. New immigrants identified strongly with the country and were proud to become citizens *because* it did not demand full cultural assimilation. A more capacious concept of citizenship absorbed ethnic attachment rather than excluded it.

The experience of African-Americans is a case apart. The racial identity of "the Negro" had been invented and imposed on slaves by their enslavers, and then used as a criterion to exclude their progeny from political citizenship and from full membership in civil society. A black child was born with the unmistakable mark of Cain. However, this imposition of a specious "identity" on blacks provoked strong identification within the African-American community itself, based on a shared history of suffering and humiliation—and of resistance, resilience, and achievement. This makes deep emotional sense. So much so that it is hard to imagine how the victims of the racial crime could ever identify as citizens of the country that had committed and justified it for centuries. America had offered protection to white religious and ethnic groups; it had enslaved the African.

It is easier to understand why there have periodically been black thinkers promoting separation, returning to Africa, leaving for cosmopolitan Europe, joining the struggle of colonized people around the globe, or overthrowing the American system. And why there is also a literature by writers who experimented with these alternatives, only to discover that they were Americans after all. But how to identify with the country emotionally, to the point that you would make sacrifices for it? How to identify if you already had made sacrifices, as black veterans returning to Jim Crow America after the Second World War had? The civil rights movement offered a constructive way of serving both the African-American community and the country as a whole: by working to force America to live up to its principles. Not just to secure formal rights, but to secure equal dignity in society as well. The leaders of the civil rights movement chose to take the concept of universal, equal citizenship more seriously than white America ever had. Not to idealize or deny difference—which was evident to the naked eye—but to render it politically impotent.

As we know, the civil rights movement then provided the template for subsequent movements to secure rights for women, homosexuals, and other groups. The parallels were hardly exact, to say the least, and there is still lingering resentment among African-Americans against whites who seem determined to draw them into a victimhood Olympics. But there is another, deeper difference between this older movement and the newer ones. In a sense, the civil rights movement had more in common with the struggles of earlier religious and ethnic minority groups, which were about having their equality and dignity as citizens recognized. This was also true of first- and second-wave feminism and the early gay rights movement. But during the 1970s and 1980s a shift began. The focus of attention was now less on the relation between our identification with the United States as democratic citizens and our identification with different social groups within it. Citizenship dropped out of the picture. And people began to speak instead of their personal identities in terms of the inner homunculus, a unique little thing composed of parts tinted by race, sex, and gender. JFK's challenge, *What can I do for my country?*—which had inspired the early sixties generation—became unintelligible. The only meaningful question became a deeply personal one: what does my country owe me by virtue of my identity?

From We to Me

A mong the more memorable romantic slogans of the 1960s was *The personal is political.* It expressed a sentiment that arises out of what romantics have always seen as the urgent need to reconcile self and world—and what anti-romantics see as an adolescent inability to live with the distinction. America has always been a fertile breeding ground for romantics, though in the first two centuries of its existence they tended to gravitate toward poetry or evangelism, whether of the Christian or the godless Emersonian kind. Political romanticism, which had roiled European politics ever since the French Revolution, was harder to find. (Which is no doubt why we gained the wholly unwarranted reputation in Europe of being a pragmatic people). The sudden outpouring of it in the early 1960s was unprecedented.

And, strangely, this romanticism had its roots in the same time and place as Reaganism: the affluent new suburbs of the 1950s. We live with two idealized images of that world. In one, favored by the right, good-paying jobs and modern technology gave Americans unprecedented prosperity and well-being; men commuted to work, women puttered around the house, and children in cowboy hats pretended to murder one another. A good time was had by all. The other image, favored by the left, is that of an air-conditioned nightmare in which men commuted to work (and drank too much), women puttered around the house (and popped pills), and children in cowboy hats pretended to murder one another (transferring their hatred of their parents onto their playmates). These are politically useful myths, nothing more.

But the darker image does capture an important truth about the time that the other ignores, a psychological truth. One has only to glance at the books Americans were reading and the movies they were watching back then to see how anxious they had become about just what kind of lives they were building for themselves out on the suburban frontier. A whole new vocabulary was developed to express that anxiety. People read about being submerged in *mass society*, of becoming an *organization man*, a faceless member of the *lonely crowd* condemned to join the *rat race*. Psychologists began conducting studies of *alienated youth*, out of concern that they were becoming aimless *juvenile delinquents*. Movie directors made films that reflected and no doubt intensified the dissatisfactions of *the man in the gray flannel suit*, *the prisoner of Second Avenue*, and the young *rebel without a cause*. Accounts of women's lives stifled by *the feminine mystique* were slower to come, but eventually they did.

It was the age of the *identity crisis*, a term coined in the early 1950s by the German psychologist Erik Erikson to describe a condition he found widespread in his prosperous adoptive country. "While the patient of early psychoanalysis suffered most under inhibitions which prevented him from being what and who he thought he already knew he was," Erikson wrote, "the patient of today suffers most under the problem of what he should believe in and who he should—or, indeed, might—become." This was hardly news; Tocqueville had rendered the same diagnosis of the American mind in the early nineteenth century. But Erikson's restatement in terms of "identity" caught the public's imagination and seemed to reflect people's own inner experience. The more the frontier settlers freed themselves from economic and social necessity, the more confused they became about what to do with their liberty. What would a meaningful, authentic life look like, now that it was possible? This question was most pressing for young people who had known only peace and prosperity. Not all those college students in bobby socks and brush cuts were surfing during spring break. Many were reading the recently translated French existentialists, Kafka's stories, the meditations of Thomas Merton, and the plays of Samuel Beckett and Eugène Ionesco, now available in cheap paperbacks. They were also joining unconventional religious groups like Campus Crusades for Christ and later the Catholic Charismatic Renewal. While their parents were absorbed in building personal wealth, they were asking themselves what it meant to be a person at all. It was this generation that made the 1960s happen.

Political romanticism is easy to spot but difficult to define. It is more a mood than a set of ideas, a sensibility that colors the way people think about themselves and their relation to society. Romantics see society itself as somewhat dubious, an imposed artifice that alienates the individual self from itself, drawing arbitrary lines, creating enclosures, and forcing us into costumes that are not of our own making. ("Society everywhere is in conspiracy against the manhood of every one of its members," wrote the tiresome Emerson.) It makes us forget who we are and inhibits us from exploring what we might become. What romantics seek is more difficult to define or articulate. Its names are legion: authenticity, transparency, spontaneity, wholeness, liberation. That the world be as one. And when the world politely declines this request, the romantic is left torn between opposed impulses. There is the impulse to flee so as to remain an authentic, autonomous self; and there is the impulse to transform society so that it seems like an extension of the self. The romantic wants to create a world where he or she will possess a fully integrated, unconflicted identity—where the answers to the questions *Who am I?* and *What are we?* are exactly the same.

When this romantic sensibility took political form in the early 1960s, older liberals and socialists couldn't for the life of them understand what young people were going on about. Civil rights, the Vietnam War, disarmament, poverty, colonialism—these were political issues certainly worth protesting about. But what did all that have to do with sassing your parents, taking drugs, listening to loud music, free love, vegetarianism, and Eastern mysticism? Yes, capitalism was the enemy of the people. But was the comb really the enemy of the soul? To an older generation the rhetoric of the time was a hopeless hash of the personal, the cultural, and the political. Trivial incidents—the canceling of a speech, the building of a school gymnasium—would let loose gushers of moral outrage, directed not at, say, Chase Manhattan Bank but at the university. The rambling Port Huron Statement published by Students for a Democratic Society (SDS) in 1962 made a number of coherent observations about foreign and domestic policy. But they were mixed in with declarations like the following:

> *The goal of man and society should be human independence: a concern not with image of popularity but with finding a meaning in life that is personally authentic; a quality of mind not compulsively driven by a sense of powerlessness, nor one which unthinkingly adopts status values, nor one which represses all threats to its habits, but one which has full, spontaneous access to present and past experiences, one which easily unites the fragmented parts of personal history, one which openly faces problems which are troubling and unresolved; one with an intuitive awareness of possibilities, an active sense of curiosity, an ability and willingness to learn.*
>
> *This kind of independence does not mean egoistic individualism—the object is not to have one's way so much as it is to have a way that is one's own.*

An inspiring passage about an inner search for meaning. But what did it have to do with voting rights in Mississippi or strikes against U.S. Steel?

To young people swept up into the New Left this all made perfect sense because, as all romantics know, *everything connects*. It stood to reason there could be no narrowly political aims divorced from the struggles for freedom and justice and authenticity in every aspect of our lives: sex relations, the family, the secretarial pool, schools, the grocery store. And around the world, too. Oppression was polymorphous and so resistance had to be, too. That was why marching in a demonstration against the Vietnam War in the morning, working at a food co-op in the afternoon, attending a feminist workshop in the evening, and then camping out on the land to get my soul free was all completely coherent. It was politics of the highest and most urgent sort. What were midterm congressional elections by comparison?

The New Left originally interpreted the slogan *The personal is political* in a somewhat Marxist fashion to mean that everything that seems personal is in fact political, that there are no spheres of life exempt from the struggle for power. That is what made it so radical, electrifying sympathizers and terrifying everyone else. But the phrase could also be taken in just the opposite sense: that what we think of as political action is in fact nothing but personal activity, an expression of me and how I define myself. As we would say today, it is a reflection of my identity. At first, the tension between the two interpretations of the slogan was not apparent to those swept up in the passions of the moment. *Legal abortion, equal wages, and day care affect me personally as a woman, but they also affect all other women. That's not narcissism; that's motivation.* But with time the tension became all too apparent, dooming the short-term prospects of the New Left and, eventually, those of American liberalism as well.

The New Left was torn apart by all the intellectual and personal dynamics that plague every left, plus a new one: identity. Racial divisions were quick to develop. Blacks complained that most leaders were white, which was true. Feminists complained that most all were men, which was also true. Soon black women were complaining about both the sexism of radical black men and the implicit racism of white feminists—who themselves were being criticized by lesbians for presuming the naturalness of the heterosexual family. What all these groups wanted from politics was more than social justice and an end to the war, though they did want that. They also wanted there to be no space between what they felt inside and what they did out in the world. They wanted to feel at one with political movements that mirrored how they understood and defined themselves as individuals. And they wanted their self-definition to be recognized. The socialist movement had neither promised nor delivered recognition: it divided the world into exploiting capitalists and exploited workers of every background. Nor did Cold War liberalism, which worked for equal rights and equal social protections for all. And certainly no recognition of personal or group identity was coming from the Democratic Party, which at the time was dominated by racist Dixiecrats and white union officials of questionable rectitude.

By the mid-1970s the New Left was gone from the national scene but was still active in community organizing in large cities like Newark, Chicago, and Oakland. (And small ones like Burlington, Vermont.) Otherwise what remained were movements and more movements that operated largely outside the Democratic Party and other political institutions. The consequences of this migration out of the party were large. The forces at work in healthy party politics are centripetal; they encourage factions and interests to come together to work out common goals and strategies. They oblige everyone to think, or at least speak, about the common good. In movement politics, the forces are all centrifugal, encouraging splits into smaller and smaller factions obsessed with single issues and practicing rituals of ideological one-upmanship. So the New Left's legacy to liberalism was a double one. It spawned issue-based movements that helped to bring about progressive change in a number of areas, most notably the environment and human rights abroad. And it spawned identity-based social movements—for affirmative action and diversity, feminism, gay liberation—that have made this country a more tolerant, more just, and more inclusive place than it was fifty years ago.

What the New Left did not do was contribute to the

unification of the Democratic Party and the development of a liberal vision of Americans' shared future. And as interest slowly shifted from issue-based movements to identity-based ones, the focus of American liberalism also shifted from commonality to difference. And what replaced a broad political vision was a pseudo-political and distinctly American rhetoric of the feeling self and its struggle for recognition. Which turned out to be not all that different from Reagan's anti-political rhetoric of the producing self and its struggle for profit. Just less sentimental and more sanctimonious.

A Primer in Pseudo-Politics

Flash forward to 1980. Ronald Reagan has been elected and Republican activists are setting out on the road to spread the new individualist gospel of small government and to campaign in out-of-the-way county, state, and congressional elections. Also on the road, though taking a different exit off the interstate, you see former New Left activists in rusting, multicolored VW buses. Having failed to overturn capitalism and the military-industrial complex, they are heading for college towns all over America, where they hope to practice a very different sort of politics within educational institutions. Both groups were successful and both left their mark on the country.

The retreat of the post-1960s left was strategic. Already in 1962 the authors of the Port Huron Statement argued that, given the power of the Dixiecrats in the Democratic Party and the quiescence of the labor movement, "we believe that the universities are an overlooked seat of influence." Universities were no longer isolated preserves of learning. They had become central to American economic life, serving as conduits and accrediting institutions for post-industrial occupations, and to political life, through research and the formation of party elites, eventually displacing unions in both spheres. The SDS authors made the case that a New Left should first try to form itself within the university, where they were free to argue among themselves and work out a more ambitious political strategy, recruiting followers along the way. The ultimate point, though, was to enter the wider world, looking "outwards to the less exotic but more lasting struggles for justice."

But as hopes for a radical transformation of American life faded, ambitions shrank. Many who returned to campus invested their energies in making their sleepy college towns into morally pure, socially progressive, and environmentally self-sustaining communities. *If we can make it there, we can make it anywhere.* Children were taken out of public schools to become test subjects in alternative educational schemes. Interminable city council meetings ended in rancor over the most radical position to take on recycling. Sister cities were sought out in Latin America, Africa, and the Middle East (though not in the conservative rural communities nearby that one passed on the way to the airport). And indeed these campus towns still do stand out from the rest of America and are very pleasant places to live, though they have lost much of their utopian allure. Most have become meccas of a new consumerist culture for the highly educated, surrounded by techie office parks and increasingly expensive homes. They are places where you can visit a bookshop, see a foreign movie, pick up vitamins and candles, have a decent meal followed by an espresso, and perhaps attend a workshop and have your conscience cleaned. A thoroughly bourgeois setting without a trace of the *demos*, apart from the homeless men and women who flock there and whose job is to keep it real for the residents.

That's the comic side of the story. The other side—heroic or tragic, depending on your politics—concerns how the retreating New Left turned the university into a pseudo-political theater for the staging of operas and melodramas. This has generated enormous controversy about tenured radicals, the culture wars, political correctness—and with good reason. But these developments mask a quieter and far more significant one. The big story is not that professors successfully indoctrinated millions of students with anti-establishment left-wing dogmas. Many certainly tried, but that seems not to have slowed the line of graduates shoving their way toward professional schools and then moving on to conventional careers. The real story is that the sixties generation passed on to students a particular conception of *what politics is*, based on its own idiosyncratic historical experience.

That experience had taught it two lessons. The first was that political activity must have some authentic meaning for the self, that one must avoid at all cost becoming just a cog in a larger machine. That is precisely what the sixties generation was fleeing from, the world of their *organization man* fathers. The second lesson, based on frustration with the immobility of American political parties and institutions, was that movement politics was the only mode of engagement that actually changes things. The lesson of these two lessons, so to speak, was that if you want to be a political person you should begin, not by joining a party, but by searching for a movement that has some deep personal meaning for you. In the 1950s and early 1960s there were already a number of such movements concerned about nuclear disarmament, war, poverty, the environment. Engaging with those issues still meant, though, having to engage with the wider world and gain some knowledge of economics, sociology, psychology, science, and especially history.

→►◄←

Pseudo-Politics

With the rise of identity consciousness, engagement in issue-based movements began to diminish somewhat and the conviction got rooted that the movements most meaningful to the self are, unsurprisingly, about the self. As the feminist authors of the Combahee River Collective put it baldly in their influential 1977 manifesto, "the most profound and potentially most radical politics come directly out of our own identity, as opposed to working to end somebody else's oppression." This new attitude had a profound impact on American universities. Marxism, with its concern for the fate of workers of the world, all of them, gradually lost its allure. The study of identity groups now seemed the most urgent scholarly/political task, and soon there was an extraordinary proliferation of departments, research centers, and professorial chairs devoted to it. In part this has been a very good thing. It has encouraged academic disciplines to widen the scope of their investigations to incorporate the experiences of large groups that had been somewhat invisible, like women and African-Americans. But it also has encouraged an obsessive fascination with the margins of society, so much so that students have come away with a distorted picture of history and of their country in the present—a significant handicap at a time when American liberals need to learn more, not less, about the vast middle of the country.

Imagine a young student entering such an environment today—not your average student pursuing a career, but a recognizable campus type drawn to political questions. She is at the age when the quest for meaning begins and in a place where her curiosity could be directed outward toward the larger world she will have to find a place in. Instead, she finds that she is being encouraged to plumb mainly herself, which seems an easier exercise. (Little does she know. . . .) She will first be taught that understanding herself depends on exploring the different aspects of her identity, something she now discovers she has. An identity which, she also learns, has already been largely shaped for her by various social and political forces. This is an important lesson, from which she is likely to draw the conclusion that the aim of education is not to progressively become a self through engagement with the wider world. Rather, one engages with the world and particularly politics for the limited aim of understanding and affirming what one already is.

And so she begins. She takes classes where she reads

histories of the movements related to whatever she decides her identity is, and reads authors who share that identity. (Given that this is also an age of sexual exploration, gender studies will hold a particular attraction.) In these courses she also discovers a surprising and heartening fact: that although she may come from a comfortable, middle-class background, her identity confers on her the status of one of history's victims. This discovery may then inspire her to join a campus group that engages in movement work. The line between self-analysis and political action is now fully blurred. Her political interest will be real but circumscribed by the confines of her self-definition. Issues that penetrate those confines now take on looming importance and her position on them quickly becomes nonnegotiable; those issues that don't touch on her identity are not even perceived. Nor are the people affected by them.

The more our student gets into the campus identity mind-set, the more distrustful she will become of the word *we*, a term her teachers have told her is a universalist ruse used to cover up group differences and maintain the dominance of the privileged. And if she gets deeper into "identity theory" she'll even start to question the reality of the groups to which she thinks she belongs. The intricacies of this pseudo-discipline are only of academic interest. But where it has left our student is of great political interest.

An earlier generation of young women, for example, might have learned that women as a group have a distinct perspective that deserves to be recognized and cultivated, and have distinct needs that society must address. Today the theoretically adept are likely to be taught, to the consternation of older feminists, that one cannot generalize about women since their experiences are radically different, depending on their race, sexual preference, class, physical abilities, life experiences, and so on. More generally, they will be taught that nothing about gender identity is fixed, that it is all infinitely malleable. This is either because, on the French view, the self is nothing, just the trace left by the interaction of invisible, tasteless, odorless forces of "power" that determine everything in the flux of life; or, on the all-American view, because the self is whatever we damn well say it is. (The most advanced thinkers hold both views at once.) A whole scholastic vocabulary has been developed to express these notions: fluidity, hybridity, intersectionality, performativity, transgressivity, and more. Anyone familiar with medieval scholastic disputes over the mystery of the Holy Trinity—the original identity problem—will feel right at home.

What matters about these academic trends is that they give an intellectual patina to the radical individualism that virtually everything else in our society encourages. If our young student accepts the mystical idea that anonymous forces of power shape everything in life, she will be perfectly justified in withdrawing from democratic politics and casting an ironic eye on it. If, as is more likely, she accepts the all-American idea that her unique identity is something she gets to construct and change as the fancy strikes her, she can hardly be expected to have an enduring political attachment to others, and certainly cannot be expected to hear the call of duty toward them.

Instead she will find herself in the hold of what might be called the Facebook model of identity: the self as a homepage I construct like a personal brand, linked to others through associations I can "like" and "unlike" at will. Citizenship, the central concept of democratic politics, is a bond linking all members of a political society over time, regardless of their individual characteristics, giving them both rights and duties. We are generally born with this status, but, through democratic political activity, we can change how it is defined and what it means. In the Facebook model of the self, the bonds that matter to me and that I decide to affirm are not political in this democratic sense. They are mere elective affinities. I can even *self-identify* with a group I don't objectively seem to belong to. In 2015 a troubled woman who was then serving as president of a local chapter of the NAACP, and who had claimed to be the victim of several anti-black hate crimes, was revealed by her parents to be, in fact, white. Her critics were outraged and the right-wing media used the episode as one more example of the loony left. But if the Facebook model of identity is correct, her supporters, and there were some, were right to defend her. If all identification is legitimately self-identification, there is no reason why this woman could not claim to be anything she imagined herself to be. And to drop that identification the moment it became too burdensome, or just a bore. Whatever.

Pseudo-Politics

→→-←←

The Facebook model of identity has also inspired a Facebook model of political engagement. During the Roosevelt Dispensation group identity became recognized not only as a legitimate way to mobilize people for political action as citizens, but also as a necessary tool for forcing our political system to fulfill its promise of equal membership. But the Facebook model is all about the self, my *very* self, not about common histories or the common good or even ideas. Young people on the left—in contrast with those on the right—are less likely today to connect their engagements to a set of political ideas. They are much more likely to say that they are engaged in politics *as an X*, concerned about other *Xs* and those issues touching on *X-ness*. They may have some sympathy for and recognize the strategic need to build alliances with *Ys* and *Zs*. But since everyone's identity is fluid and has multiple dimensions, each deserving of recognition, alliances will never be more than marriages of convenience.

The Once and Future Liberal

→>◄←

The more obsessed with personal identity campus liberals become, the less willing they become to engage in reasoned political debate. Over the past decade a new, and very revealing, locution has drifted from our universities into the media mainstream: *Speaking as an X . . .* This is not an anodyne phrase. It tells the listener that I am speaking from a privileged position on this matter. (One never says, *Speaking as a gay Asian, I feel incompetent to judge this matter.*) It sets up a wall against questions, which by definition come from a *non-X* perspective. And it turns the encounter into a power relation: the winner of the argument will be whoever has invoked the morally superior identity and expressed the most outrage at being questioned. So classroom conversations that once might have begun, *I think A, and here is my argument*, now take the form, *Speaking as an X, I am offended that you claim B.* This makes perfect sense if you believe that identity determines everything. It means that there is no impartial space for dialogue. White men have one "epistemology," black women have another. So what remains to be said?

Pseudo-Politics

What replaces argument, then, is taboo. At times our more privileged campuses can seem stuck in the world of archaic religion. Only those with an approved identity status are, like shamans, allowed to speak on certain matters. Particular groups—today the transgendered—are given temporary totemic significance. Scapegoats—today conservative political speakers—are duly designated and run off campus in a purging ritual. Propositions become pure or impure, not true or false. And not only propositions but simple words. Left identitarians who think of themselves as radical creatures, contesting this and transgressing that, have become like buttoned-up Protestant schoolmarms when it comes to the English language, parsing every conversation for immodest locutions and rapping the knuckles of those who inadvertently use them.

A strange and depressing development for professors who went to college back in the 1960s, rebelled against the knuckle rappers, and mussed the schoolmarm's hair. Things seem to have come full circle: now the students are the narcs. That was hardly the intention when the New Left, fresh from real political battles in the great *out there*, returned to campus in hopes of encouraging the young to follow in their footsteps. They imagined raucous, no-holds-barred debates over big ideas, not a roomful of students looking suspiciously at one another. They imagined being provocative and forcing students to defend their positions, not getting emails from deans suggesting they come in for a little talk. They imagined launching their politically committed and informed students into the world, not watching them retreat into themselves. What happened?

Another Word from Karl Marx

A serious Marxist—there still are some—would not have been surprised. Marxism as an ideology had many faults, but at least one large virtue: it forced those who adhered to it to look up from their particular situations and engage intellectually with the deep forces that shape history—forces like class, war, religion, and science. (It had trouble with race, which it tended to collapse into a matter of class.) Marxists kept their eyes focused on the horizon; often they saw things upside down, or saw chimeras, but at least they were looking. With the rise of liberal identity consciousness, all eyes have turned within. As many progressives have complained, and rightly so, the rhetoric of identity has crowded out the analysis of class and how it has changed with our new economy. Not too long ago liberal politics aimed to inspire individuals to actively remake society. Today's focus is on the passive social construction of individuals.

A Marxist analysis of this transformation might go like this: The election of Ronald Reagan marked a new stage in the history of advanced capitalism. The politics of the post–Second World War period were shaped by liberal and progressive efforts to blunt the most egregious effects of capitalism by building the welfare state, strengthening regulations, instituting protections for workers, and fighting for the enfranchisement of African-Americans. Valiant as these efforts were, they did not get to the root cause of the problem, which was capitalism itself. These reform efforts instead became associated *with* capitalism, not with efforts to destroy it. And so, when the oil crisis of the mid-1970s threatened the economic growth that post-war America had come to expect, the country turned, not against corporations and banks, but against liberalism. Thanks are due almost entirely to a perverse right-wing ideology that convinced people that the cure to the ills of capitalism is . . . more capitalism. Less solidarity, more individualism. Less charity, more greed. Less politics, more family and self.

It is hardly a coincidence, such a Marxist might continue, that a cult of personal identity also developed in our universities in the age of Reagan and became the governing ideology of the liberal power elite in the Democratic party, the media, and the education and legal professions. While many students studied business and economics in order to make money for themselves, others were taking classes where they learned how very special those selves are. Some took both sorts of classes, satisfying both their pocketbooks and their consciences. The intellectual and material forces of the age were working together to keep them self-involved, and to convince them that narcissism with attitude was both good business and good politics. Identity is not the future of the left. It is not a force hostile to neoliberalism. Identity is Reaganism for lefties.

III

Politics

Politics is slow, steady drilling through hard boards.

—Max Weber

The first effort, then, should be to state a vision.

—"The Port Huron Statement" (1962)

Reset

And so concludes our tale of anti-politics and pseudo-politics in the long age of Reagan. Now what can liberals learn from it?

The most important lesson is this: that for two generations America has been without a *political* vision of its destiny. There is no conservative one; there is no liberal one. There are just two tired individualistic ideologies intrinsically incapable of discerning the common good and drawing the country together to secure it under present circumstances. We are governed by parties that no longer know what they want in a large sense, only what they don't want in a small sense. Republicans don't want the programs and reforms that are the legacy of the New Deal, the New Frontier, and the Great Society. Democrats don't want Republicans to cut them. But what are the parties' ultimate aims, whatever the size and shape of government? What are they fundamentally after? What sort of image of the future governs their actions? They seem no longer to know. So the public can hardly be expected to. We find ourselves in a post-vision America.

It is difficult to discuss political vision without sounding faintly ridiculous. It is not something you can shop for. You can't grow it, mine it, or hunt for it. There are no laboratories for discovering it, no candidates with résumés lined up to be interviewed for the position. Political vision emerges of its own accord out of the timely encounter of a new social reality, ideas that capture this reality, and leaders capable of linking idea and reality in the public mind so that people feel the connection. (Understanding it is less important.) The advent of leaders blessed with that gift, like Roosevelt and JFK and Reagan, is as impossible to predict as the return of the Messiah. All we can do is prepare.

How we react to the presidency of Donald Trump will be the first test of our preparedness. His administration in its infancy is already wracked with scandals. But the real scandal is that he is president at all. Yes, a few extra votes in key states might have changed the electoral college outcome. But a Democratic victory would not have masked the fact that it was a third force that surged from below to fill a vacuum and defeat both parties. There proved to be an untapped yearning to hear someone address America's new challenges in a different key, someone willing to champion change and say without equivocation that America can be great. Trump offered an authoritarian snarl and an ever-changing string of bizarre spontaneous "positions," not a political vision. But his demagogic skills were sufficient to move millions to applaud his race baiting, his misogyny, his hardly veiled threats of violence, his contempt for the press, and his contempt for the law.

The effects can already be felt: with each passing day our public life is getting uglier. So it is encouraging to see how quickly liberals have organized to resist Trump. But resistance is by nature reactive; it is not forward-looking. And anti-Trumpism is not a politics.

The Once and Future Liberal

My worry is that liberals will get so caught up in countering his every move, essentially playing his game, that they will fail to seize—or even recognize—the opportunity he has given them. Now that he has destroyed conventional Republicanism and what was left of principled conservatism, the playing field is empty. For the first time in living memory, we liberals have no ideological adversary worthy of the name. So it is crucial that we look beyond Trump.

The only adversary left is ourselves. And we have mastered the art of self-sabotage. At a time when we liberals need to speak in a way that convinces people from very different walks of life, in every part of the country, that they share a common destiny and need to stand together, our rhetoric encourages self-righteous narcissism. At a moment when political consciousness and strategizing need to be developed, we are expending our energies on symbolic dramas over identity. At a time when it is crucial to direct our efforts into seizing institutional power by winning elections, we dissipate them in expressive movements indifferent to the effects they may have on the voting public. In an age when we need to educate young people to think of themselves as citizens with duties toward each other, we encourage them instead to descend into the rabbit hole of the self. The frustrating truth is that we have no political vision to offer the nation, and we are thinking and speaking and acting in ways guaranteed to prevent one from emerging.

In the aftermath of the collapse of the Soviet Union, and with it whatever hopes were still invested in Communism, a group of reform-minded Italian leftists started a lively political magazine titled *Reset*. (It has since morphed into a lively website.) The title was well chosen and reflected the editors' conviction that a certain idea of the left, a certain tradition of thought and action, had been defeated in no uncertain terms. And so it was time to rethink fundamental assumptions, question old dogmas, break bad habits, and puncture taboos.

With the election of Donald Trump, American

liberalism has reached its *Reset* moment. It is time to reacquaint ourselves with the demands, the possibilities, and the constraints of democratic politics in our system. As a small contribution to this effort, I conclude with some lessons that can be drawn from the history and analysis I have offered.

The first three have to do with priorities: the priority of institutional over movement politics; the priority of democratic persuasion over aimless self-expression; and the priority of citizenship over group or personal identity. The fourth has to do with the urgent need for civic education in an increasingly individualistic and atomized nation. Others might draw different lessons from my story, or add their own, or question the story itself. That's fine. The point is to start focusing attention on whatever barriers we have erected between us and the American public, and between us and the future. And we must begin by questioning the taboos—particularly the taboos surrounding identity—that have protected those barriers from scrutiny. Our common goal must be to put ourselves in a position to develop an inspiring, optimistic vision of what America is and what it can become through liberal political action.

The Marcher and the Mayor

During the Roosevelt Dispensation the two grand themes of American liberalism were justice and solidarity. And liberals understood what should still be obvious today, which is that securing those things depends ultimately on acquiring and holding power in established democratic institutions—executive offices, legislatures, courts, and bureaucracies. As a corollary it followed that, given our federalized system, winning elections across the country was the first order of political business.

Identity liberals, though, have never absorbed this lesson. They remain under the spell of movement politics. The role of social movements in American history, while important, has been seriously inflated by left-leaning activists and historians, though for an understandable reason: between the 1950s and the 1980s the country was indeed transformed by organized efforts to secure the rights of African-Americans, women, and gay Americans. There is, though, a natural tendency for anyone who lives through such a transformation to begin reading the past in light of it, and then to project an imagined trend outward into the future. The result is Whiggish history. Many if not most of the first-year students I find in my college classes have been taught American history in this way or have picked it up from films and television documentaries. The spine, so to speak, of their skeleton knowledge of American history is a narrative that moves smoothly from abolition and the women's suffrage movements in the nineteenth century, to the labor movements of the early twentieth century, and ends with the more recent movements I've mentioned. It seems to have left them with the strong impression that a historical process has been unfolding and is destined to continue into the future.

The problem with this narrative is that it gives a false impression of what the main focal point of American democratic politics is and always has been: government. The framers of our Constitution arranged things so that political action would have to be filtered through institutions that require consultation and compromise, and would depend on a system of frequent elections, checks and balances, the autonomy of the civil service, civilian control of the military, the writing of laws and regulations, and their impartial enforcement. And all this would have to be done at three levels of government. This meant that being politically successful would require a lot of tedious, incremental work, which for the framers was a recommendation. They wanted to spare the United States the fate of Europe, which they saw as wracked for centuries by the arbitrary rule of tyrants, court intrigues, coups d'état, wars of religion, and republican factionalism. The stuff of poetry, but stifling to the human spirit. How much better, they thought, to canalize political energy into institutions, while making them as transparent and participatory as possible.

Romantics chafe at this undramatic conception of politics. They prefer to think of it as a zero-sum confrontation—the People against Power, or Civilization against the Mob. And it's not hard to see why. What could be more stirring than history seen as a series of revolutions, counterrevolutions, restorations, manifestos, mass marches, dissidents, police repression, general strikes, arrests, jailbreaks, anarchist bombings, and assassinations? And what could be more dreary than the history of parties and public administration and treaties? There was a strong antiliberal streak in European political thinking running from the French Revolution until quite recently, on the left and right, and it was inspired as much by aesthetic disdain for democratic dullness as by moral conviction.

When Marxian socialism came to the United States after the 1848 revolutions, it brought along in its baggage this European suspicion of liberal-democratic procedures. Eventually that was dispelled and socialist organizations began participating in electoral politics. But they continued to think of themselves more as the vanguard of a movement than as voices in a democratic chorus. And their preferred political tactics remained the mass demonstration and the strike—rather than, say, winning elections for county commissioner. The significance of these groups in American politics peaked during the Great Depression and then faded. But their movement ideal retained its grip on the left, and in the 1960s it captured the imagination of liberals as well. There had been emancipatory movements before, against slavery, for women's rights, for workers' protection. They did not question the legitimacy of the American system; they just wanted it to live up to its principles and respect its procedures. And they worked with parties and through institutions to achieve their ends. But as the 1970s flowed into the 1980s, movement politics began to be seen by many liberals as an alternative rather than a supplement to institutional politics, and by some as being more legitimate. That's when what we now call the social justice warrior was born, a social type with quixotic features whose self-image depends on being unstained by compromise and above trafficking in mere interests.

Yet it is an iron law in democracies that anything achieved through movement politics can be undone through institutional politics. The reverse is not the case. The movements that reshaped our country over the last half century did much good, especially in changing, as we say, hearts and minds. That is perhaps the most important thing any movement does, as Gandhi and Martin Luther King Jr. believed. But over the long term they are incapable of achieving concrete political ends on their own. They need system politicians and public officials sympathetic to movement aims but willing to engage in the slow, patient work of campaigning for office, drawing up legislation, making trades to get it passed, and then overseeing bureaucracies to see that it is enforced. Martin Luther King Jr. was the greatest movement leader in American history. But, as Hillary Clinton once correctly pointed out, his efforts would have been futile without those of the machine politician Lyndon Johnson, a seasoned congressional deal maker willing to sign any pact with the devil to get the Civil Rights Act and Voting Rights Act passed.

And the work doesn't stop once legislation is passed. One must keep winning elections to defend the gains that social movements have contributed to. If the steady advance of a radicalized Republican Party, over many years and in every branch and at every level of government, should teach liberals anything, it is the absolute priority of winning elections today. Given the Republicans' rage for destruction, it is the *only* way to guarantee that newly won protections for African-Americans, other minorities, women, and gay Americans remain in place. Workshops and university seminars will not do it. Online mobilizing and flash mobs will not do it. Protesting, acting up, and acting out will not do it. The age of movement politics is over, at least for now. We need no more marchers. We need more mayors. And governors, and state legislators, and members of Congress . . .

Demos, as in Democracy, as in Democrat

I have called the history of identity liberalism the story of an abdication. And I have portrayed it as a turn toward the self. But it was also a turn away from contact with much of the country and many of the people whose views are not exactly our own on every issue. By this I do not mean to suggest that there is some hidden, homogeneous "silent majority" or "real America" whose views are more important or virtuous than those of others and must be genuflected to. That was the founding myth of the new American right, beginning with Barry Goldwater, and the 2016 election buried whatever remained of it. I mean rather that by getting so focused on themselves and the groups they felt they belonged to, identity liberals acquired additional disdain for ordinary democratic politics because it meant engaging with and persuading people unlike themselves. Instead they began delivering sermons to the unwashed from a raised pulpit.

This detachment from the *demos* was even institutionalized. After the 1968 convention, Democratic Party rules were significantly rewritten, ostensibly to open the party to groups and interests that had been ignored by the unions and city bosses. As most historians agree, the unintended consequence was to marginalize the blue-collar unions and public officials who had been the pillars of the party structure, and replace them with educated activists tied to single issues or to particular presidential campaigns. This robbed the party of people who could easily talk and listen to both educated liberal elites and the Democrats' voting base. The kind of people who were adept at measuring the temperature and barometric pressure of public opinion, keeping the elites informed about the political weather outside and suggesting when to take an umbrella.

→>-<←

Distrust of the legislative process and increased reliance on the courts to achieve their goals also detached liberal Democratic elites from a wider base. To pass legislation you need to persuade very different sorts of people that it makes sense, which might require compromise but also helps ensure that the law will not provoke a mass reaction that leaves you in a worse position than when you began. Legislation can be tweaked, and negotiations about it are usually about how to balance a number of relative goods. In ordinary democratic politics, groups represent interests that can be defended but also balanced against each other when necessary to get agreement. To get standing in court, on the other hand, all you have to do is present your case as a matter of absolute legal right, and the only people you have to persuade are the judges assigned to your case.

This was an essential tactic in the early years of the civil rights movement, but it has been a disaster for liberalism's reputation with the public ever since. It got liberals into the habit of treating every issue as one of inviolable right, leaving no room for negotiation, and inevitably cast opponents as immoral monsters, rather than simply as fellow citizens with different views. And it also relieved liberals of the patient work of finding out where people stand, trying to persuade them, and building a social consensus, which is the most secure foundation for any social policy. Liberals' legalistic approach created a large opening for the Republicans to claim that they were the true representatives of the *demos*, while the Democrats represented a caste of high priests. And the image remains in the public mind.

Americans are a strange breed. We love to preach, and we hate being preached at. In one hemisphere of our brains the sermons of Cotton Mather run on an infinite loop; in the other we hear the echo of Mark Twain's laughter. When the Twain side is napping the Mather side undergoes a Great Awakening. Surges of fevered fanaticism come over us, all sense of proportion is lost, and everything seems of an unbearable moral urgency. *Repent, America, repent now!* The country is undergoing such an Awakening at this very moment concerning race and gender, which is why the rhetoric being generated sounds evangelical rather than political. That one now hears the word *woke* everywhere is a giveaway that spiritual conversion, not political agreement, is the demand. Relentless speech surveillance, the protection of virgin ears, the inflation of venial sins into mortal ones, the banning of preachers of unclean ideas—all these campus identity follies have their precedents in American revivalist religion. Mr. Twain might have found it amusing but every opinion poll shows that the vast majority of Americans do not.

Liberals have elections to contest and centrist working-class voters to win back. That is job number one. And nothing will turn voters off more surely than being hectored in this way. So a couple of reminders to the identity conscious:

Elections are not prayer meetings, and no one is interested in your personal testimony. They are not therapy sessions or occasions to obtain recognition. They are not seminars or "teaching moments." They are not about exposing degenerates and running them out of town. If you want to save America's soul, consider becoming a minister. If you want to force people to confess their sins and convert, don a white robe and head to the River Jordan. If you are determined to bring the Last Judgment down on the United States of America, become a god. But if you want to win the country back from the right, and bring about lasting change for the people you care about, it's time to descend from the pulpit.*

And once you do descend, learn to listen and imagine. You need to visit, if only with your mind's eye, places where Wi-Fi is nonexistent, the coffee is weak, and you will have no desire to post a photo of your dinner on Instagram. And where you'll be eating with people who give genuine thanks for that dinner in prayer. Don't look down on them. As a good liberal you have learned not to do that with peasants in far-off lands; apply the lesson to Southern Pentecostals

---

* Children do not respond well to scolding and neither do nations. It just puts their backs up. They become better only when they are told that they are already good and *therefore* can improve. It would be wise for us to admit—indeed celebrate the fact—that on some of the issues we care about, things have gotten better, thanks to our efforts: the building of an educated black middle class, improved job opportunities for women, social acceptance of homosexuality, and so on. We should be using progress already made as an incentive to continue the work rather than pretending that things have never been so bad and Americans never so morally depraved.

and gun owners in the mountain states. Just as you wouldn't think of dismissing another culture's beliefs as mere ignorance, don't automatically attribute what you are hearing to the right-wing media machine (as loathsome and corrupting as it is). Try to hear what's behind the false assertions and see if you can't use it to make a connection.*

Democratic politics is about persuasion, not self-expression. *I'm here, I'm queer* will never provoke more than a pat on the head or a roll of the eyes. Accept that you will never agree with people on everything—that's to be expected in a democracy. One effect of engaging in social movements tied to identity is that you've been

---

* Electoral politics is a little like fishing. When you fish you get up early in the morning and go to where the fish are—not to where you might wish them to be. You then drop bait into the water (bait being defined as something they want to eat, not as "healthy choices"). Once the fish realize they are hooked they may resist. Let them; loosen your line. Eventually they will calm down and you can slowly reel them in, careful not to provoke them unnecessarily. The identity liberals' approach to fishing is to remain on shore, yelling at the fish about the historical wrongs visited on them by the sea, and the need for aquatic life to renounce its privilege. All in the hope that the fish will collectively confess their sins and swim to shore to be netted. If that is your approach to fishing, you had better become a vegan.

surrounded by the like-minded and like-faced and like-educated. Impose no purity tests on those you would convince. Not everything is a matter of principle—and even when something is, there are usually other, equally important principles that might have to be sacrificed to preserve this one. Moral values are not pieces in a puzzle where everything has been precut to fit.

An example. I am an absolutist on abortion. It is the social issue I most care about, and I believe it should be safe and legal virtually without condition on every square inch of American soil. But not all my fellow citizens agree (though in certain cases an enormous majority does). So what should my strategy be? Drive pro-life voters out of the garden and into the waiting arms of the radical right? Or should I find a civil way to agree to disagree and make a few compromises in order to keep the liberal ones in my own party and voting with me on other issues?

The Democratic Party first performed this experiment on itself at the 1992 national convention. That was the year Robert P. Casey, the Catholic Pennsylvania governor who had worked relentlessly to expand social services in his state and was very pro-union, asked to address the convention and present a pro-life plank to the platform, even though he knew it would be defeated. His request was denied. That sent a strong signal to working-class Catholic and evangelical voters that if they did not fall into line on this one issue they were no longer welcome in the party. In the run-up to the Women's March on Washington in January 2017 the same thing happened to religious feminist groups that wanted to express their disgust with Donald Trump but were opposed to abortion. They were disinvited. And one more bridge was burned.

Citizens, United

Identity liberalism banished the word *we* to the outer reaches of respectable political discourse. Yet there is no long-term future for liberalism without it. Historically liberals have called on *us* to ensure equal rights, they want *us* to feel a sense of solidarity with the unfortunate and help them. *We* is where everything begins. Barack Obama understood this, which is why he so often said *Yes, we can* and *That's not who we are*. (Though, characteristically, he never got around to saying who exactly we are or who we might become.) But by abandoning the word, identity liberals have landed themselves in a strategic contradiction. When speaking about themselves, they want to assert their difference and react testily to any hint that their particular experience or needs are being erased. But when they call for political action to assist their group *X*, they demand it from people they have defined as *not-X* and whose experiences cannot, they say, be compared with their own.

But if that is the case, why would these others respond? Why should *not-Xers* give a damn about *Xers*, unless they believed they share something with them? Why should we expect them to feel anything at all?

The only way out of this conundrum is to appeal to something that as Americans we all share but which has nothing to do with our identities, without denying the existence and importance of the latter. And there is something, if only liberals would again begin to speak of it: citizenship.*

Admittedly, the word *citizen* has a musty air and conjures up images, for people of a certain age, of schoolteachers tapping blackboards with wooden pointers during civics class. But it has great democratic—and Democratic—potential, especially today. That is because citizenship is a *political status*, nothing less and nothing more. To say that we are all citizens is not to say that we are all alike in every respect. It is a social fact that many Americans today think of themselves in terms of identity groups, but there is no reason why they cannot simultaneously think of themselves as political citizens like everyone else. Both ideas can be— indeed, are—true. What's crucial at this juncture in our history is to concentrate on this shared political status, not on our other manifest differences. Citizenship is a crucial weapon in the battle against Reaganite dogma because it brings home the fact that we are part of a legitimate common enterprise that *We, the people*

* To repeat: nothing I say here about citizenship should be taken to imply anything about who should be granted it or how noncitizens should be treated. I am interested here only in what citizenship *is*.

have freely willed into being. That we are not elementary particles.

Another reason to think and speak in terms of political citizenship is that the status is extendable and its meaning expandable. The American right uses the term *citizenship* today as a tool of exclusion, but liberals have traditionally seen it as a generous tool for inclusion. The modern concept of citizenship originally meant that you were neither a slave nor a subject under the authority of a monarch or a pope. It was a formal designation and was restricted to a small class of people. In the nineteenth century, in Europe and the United States, the question became who deserved to be citizens, and eventually the formal franchise was extended to those without property, women, former slaves, and so on. In the twentieth century the question centered on what was materially necessary to enjoy the benefits of democratic citizenship equally, which provided a way of making the case for the modern welfare state. All liberal arguments to improve that state can still be formulated in terms of citizen enfranchisement.

→-◄-

And the concept of citizenship has one additional advantage. It provides a political language for speaking about a solidarity that transcends identity attachments. Democratic citizenship implies reciprocal rights and duties. We have duties because we have rights; we enjoy rights because we do our duty. Yet the more individualistic our society has become and the more assertive Americans have become about their rights, the more narrowly political discussion and legal discourse have revolved around the self. In the Second World War era the connection between rights and duties didn't have to be elaborated; the fight against fascism made it obvious. It was still obvious to the young men and women who served in Korea and even to those who volunteered in the early years of the Vietnam War. But the debacle of Vietnam made the notion of duty laughable to those who opposed it, and soon to much of the country.

The political creeds of our time make it virtually impossible to discuss duty. If you *don't need the goddamn government for anything*, why should you do anything for your country? It was striking during the Reagan years that for all his talk about America as the last line of defense against tyranny, not once in his presidency did he ask the public to make any sacrifices to defend freedom at home or abroad. He understood the libertarian mood of the country too well. We now find it much easier to expand the deficit and rely on an all-volunteer force, then to give soldiers priority boarding on planes and thank them for their service. But why even do that? They got paid.

Progressives understand the need for solidarity in a way identity liberals do not—which is one reason among many why it may be up to progressives to save contemporary liberalism from itself.* But progres-

---

* Consider this statement Bernie Sanders made not long after the 2016 election:

> One of the struggles that you're going to be seeing in the Democratic Party is whether we go beyond identity politics. I think it's a step forward in America if you have an African-American head or CEO of some major corporation. But you know what, if that guy is going to be shipping jobs out of his country and exploiting his workers, doesn't mean a whole hell of a lot if he's black or white or Latino. . . . We need candidates—black and white and Latino and gay and male. We need all of that. But we need all of those candidates and public officials to have the guts to stand up to the oligarchy. That is the fight of today.

sives don't speak of duty either. They remain prisoners of their own fixation on class and their nostalgia for America's industrial union past. Progressives are completely right to argue that class matters as much now as it did in the first Gilded Age. But class consciousness has far less effect on the human mind—and certainly on the American mind—than those of a Marxist bent like to think. And if a sense of solidarity is based solely on economic resentment, it will be shared only by those who feel disadvantaged, and will disappear as soon as their fortunes improve in an economic upturn. Progressive political rhetoric does nothing to convince the well-off that they have a permanent duty to the worse-off. The Bible used to, but no longer. Though this is still a churchgoing nation, the gospel now being preached, particularly in evangelical circles, has been infected with the same individualism, selfishness, and superficiality that have infected other

Evidence of how difficult this struggle may be was inadvertently provided a few days later by Bernie's former spokeswoman Symone Sanders (no relation), who declared, "In my opinion we don't need white people leading the Democratic Party right now. The Democratic Party is diverse, and it should be reelected as so in leadership and throughout the staff, at the highest levels."

sectors of American life. Many believers still tithe to their churches but they reject outright the notion that taxes, too, are a kind of democratic tithe that goes to help fellow citizens like themselves. Charity, like tipping, is now left to the customer's discretion.

In the absence of a motivating charitable faith, the only way one can hope to induce a sense of duty is by establishing some sort of identification between the privileged and the disadvantaged. Citizenship is not an identity in the way we currently use the term, but it provides one possible way of encouraging people to identify with one another. Or at least it provides a way to talk about what they already share. There is good reason why progressives should stop framing their calls for economic justice in terms of class and start appealing instead to our shared citizenship.[*]

---

[*] American progressives' single-minded focus on economics owes more to Marxism than to the original Progressive movement. If they want to become a major force in American politics again, it would be wise for them to look back to the movement's founders and their generous view of the country and its destiny, rather than to the latest books from Verso Press. Teddy Roosevelt should be required reading for all Bernie Sanders voters today (though they will have to skip over the jingoistic bits).

→>-<-

Identity liberals should follow suit. I have been hard on them in these pages, and with just cause, for separating us more than we already are separated in our individualistic age. But the concrete issues they care about are all too real. It is unconscionable that black motorists and pedestrians have been regularly singled out by police officers who then handled them violently, with impunity in some places. It is obscene that some young men, minorities in large part, are handed long sentences for selling drugs used by the poor, while those who sell drugs to the rich serve short ones. It is undemocratic that some women receive lower pay for the same work men perform. It is wrong that in some places gay couples walking hand in hand can be threatened on the street—and transgender people, suffer far worse—without the perpetrators fearing punishment. It is shameful that for so long their partners were not accorded the basic rights and dignities that married couples enjoy. Why? Because these are our fellow citizens who deserve to be fully enfranchised. That is all any other American should need to know—and all we should have to appeal to.

The Once and Future Liberal

Equal protection under the law is not a hard principle to convince Americans of. The difficulty comes in persuading them that it has been violated in particular cases, and of the need to redress the wrong. Prejudice and indifference run deep. Education, social reform, and political action can persuade some. But most people will not feel the sufferings of others unless they feel, even in an abstract way, that *it could have been me or someone close to me*. Consider the astonishingly rapid transformation of American attitudes toward homosexuality and even gay marriage over the past decades. Gay activism brought these issues to public attention but attitudes were changed during tearful conversations over dinner tables across America when children came out to their parents (and, sometimes, parents came out to their children). Once parents began to accept their children, extended families did too, and today same-sex marriages are celebrated across the country with all the pomp and joy and absurd overspending of traditional American marriages. Race is a wholly different matter. Given the segregation in American society white families have little chance of seeing and therefore understanding the lives of black Americans. I am not a black male motorist and never will be. All the more reason, then, that I need some way to identify with one if I am going to be affected by his experience. And citizenship is the only thing I know we share. The more the differences between us are emphasized, the less likely I will be to feel outrage at his mistreatment.

Black Lives Matter is a textbook example of how not to build solidarity. There is no denying that by publicizing and protesting police mistreatment of African-Americans the movement mobilized supporters and delivered a wake-up call to every American with a conscience. But there is also no denying that the movement's decision to use this mistreatment to build a general indictment of American society, and its law enforcement institutions, and to use Mau-Mau tactics to put down dissent and demand a confession of sins and public penitence (most spectacularly in a public confrontation with Hillary Clinton, of all people), played into the hands of the Republican right.

As soon as you cast an issue exclusively in terms of identity you invite your adversary to do the same. Those who play one race card should be prepared to be trumped by another, as we saw subtly and not so subtly in the 2016 presidential election. And it just gives that adversary an additional excuse to be indifferent to you. There is a reason why the leaders of the civil rights movement did not talk about identity the way black activists do today, and it was not cowardice or a failure to be *woke*. The movement shamed America into action by consciously appealing to what we share, so that it became harder for white Americans to keep two sets of books, psychologically speaking: one for "Americans" and one for "Negroes." That those leaders did not achieve complete success does not mean that they failed, nor does it prove that a different approach is now necessary. No other approach is likely to succeed. Certainly not one that demands that white Americans agree in every case on what constitutes discrimination or racism today. In democratic politics it is suicidal to set the bar for agreement higher than necessary for winning adherents and elections.

Liberals' Education

Citizens are made, not born. Sometimes historical forces do the work. War in particular can evoke a sense of civic belonging and solidarity that didn't exist before. War can also extinguish it, as happened after the First World War in Europe and in the United States during the Vietnam War. The best one can hope for is that democratic citizens will be formed through an education in the principles of self-government. But that's only a start. For those principles to then motivate action they must be rooted in a feeling we are not born with. And feelings can't be taught; they have to be conjured up. It's the closest thing to a miracle that exists in politics.

Because sustaining civic feeling is so difficult, democracies are subject to entropy. When the bond of citizenship is badly cast or has been allowed to weaken, there is a natural tendency for subpolitical attachments to become paramount in people's minds. We see this in every failed American effort to export democracy abroad. And we are also seeing it in Eastern Europe today, a particularly tragic development. Within a few years after the Berlin Wall fell in 1989, democratic institutions were established there. But not a sense of shared citizenship, which is the work of generations. Democracies without democrats do not last. They decay, into oligarchy, theocracy, ethnic nationalism, tribalism, authoritarian one-party rule, or some combination of these.

For most of its history the United States has been lucky enough to evade these classic forces of entropy, even after a devastating Civil War and mass immigration. What's extraordinary—and appalling—about the past four decades of our history is that our politics have been dominated by two ideologies that encourage and even celebrate the *unmaking* of citizens. On the right, an ideology that questions the existence of a common good and denies our obligation to help fellow citizens, through government action if necessary. On the left, an ideology institutionalized in colleges and universities that fetishizes our individual and group attachments, applauds self-absorption, and casts a shadow of suspicion over any invocation of a universal democratic *we*. This at a time when, precisely because America has become more diverse and individualistic in reality, there is greater, not less, need to cultivate political fellow feeling.

And not only for this reason. Anyone who watched a televised Donald Trump rally in the 2016 campaign was witness to a mob orgy, not an assembly of citizens. What was most shocking about the verbal and physical violence, the menacing taunts directed at journalists, the threats against rivals (*Lock her up!*), the conspiracy mongering (*It's all rigged!*) was how unshocked those in attendance were. Many cheered on and the rest just shrugged their shoulders. *He tells it like it is*, they lamely told journalists as they left. Or they offered assurance that Trump had no intention of following through on what he said—as if that were reassuring. Whatever might be said about the legitimate concerns of Trump supporters, they have no excuse for voting for him. Given his manifest unfitness for higher office, a vote for Trump was a betrayal of citizenship, not an exercise of it.

What became perfectly evident during the campaign is that his voters were generally clueless about how our democratic institutions work, and about all the informal rules and norms that keep them working well. All they seemed to possess was a paranoid, conspiratorial picture of power that our popular culture and right-wing media continually refresh. (*Mr. Smith Goes to Washington* has as much to answer for as Fox News.) Wanting to *shake things up* and leave it at that is not a democratic impulse. It is a tyrannical one that demagogues from ancient Greece to the present have cultivated and exploited. And now that Trump actually is shaking things up—turning his administration into a bizarre hybrid of junta and family business, Tweeting unhinged claims, purging those thought insufficiently servile—how have our flag-draped Tea Party Patriots responded? By defending him.

Have we simply ceased making citizens in large parts of the country? Are we becoming one more democracy without democrats?

This should be one of our chief concerns. There are many ways in which we become Americans. We learn to believe in the uniqueness of the individual when we are young; in schools and the workplace we learn to blend in and not be "arrogant" (the cardinal sin in American eyes); in churches and volunteer organizations we learn to cooperate; and in television shows and movies and even commercials we have all these values reinforced with mind-numbing regularity. We have no problem breeding "the American" as a social type, and immigrants become Americans in this sense with astonishing speed and ease.

But in order to become naturalized citizens, immigrants must also pass a civics test. It is not easy. They are asked about the principles of democratic government and about very specific features of the U.S. Constitution and the Declaration of Independence. They are expected to know about the structure of our institutions at every level of government and the relative powers of each branch. They are asked about citizens' rights, but also about their duties. The test even contains detailed questions about American history, beginning with the nation's founding and moving forward to the present. If you know anyone who has taken the test, you know how much it meant to him or her to pass it, and how moved he or she was to pledge allegiance to the flag. The experience made them feel attached to the country that had accepted them.

Their children, like others born here, will take no such test. Instead they will find themselves thrown into a hyperindividualistic culture in which personal choice and self-definition have become idols. They will have learned next to nothing about public obligation and the need for solidarity among citizens. Worse, they will not have been raised to *feel* solidarity, except with those they have chosen to associate with. And why would they? The market economy does not instill the feeling; our schools do not instill it; our churches do not instill it; our popular culture does not instill it. (Instead it instills sentimentality, which is self-referential.) It's true: many young people develop the feeling nonetheless. Some, disproportionately from religious Southern families, join the armed forces for principled reasons. Others, disproportionately from the upper-middle class, do a stint with Teach For America or similar groups after college. Somehow these young people feel the civic bond and have a sense of duty. So we know it can still exist. The question is whether liberals will commit themselves to strengthening that bond by emphasizing what we all share and owe one another as citizens, not what differentiates us.

Of all the developments I have discussed in this book, the most self-defeating from a liberal standpoint is identity-based education. Conservatives are right: our educational institutions, from bottom to top, are mainly run by liberals, and teaching has a liberal tilt. But they are wrong to infer that students are therefore being politicized. The liberal pedagogy of our time, focused as it is on identity, is actually a depoliticizing force. It has made our children more tolerant of others than certainly my generation was, which is a very good thing. But by undermining the universal democratic *we* on which solidarity can be built, duty instilled, and action inspired, it is unmaking rather than making citizens. In the end this approach just strengthens all the atomizing forces that dominate our age.[*]

---

[*] One of the small ironies of life in the Reagan years is that conservatives, seeing that they were locked out of the university,

The Once and Future Liberal

It's strange: liberal academics idealize the sixties generation, as their weary students know. But I've never heard any of my colleagues ask an obvious question: what was the connection between that generation's activism and what they learned about our country in school and in college? After all, if professors would like to see their own students follow in the footsteps of the Greatest Generation, you would think they would try to reproduce the pedagogy of that period. But they don't. Quite the contrary. The irony is that the supposedly bland, conventional schools and colleges of the 1950s and early 1960s incubated what

created a parallel intellectual universe, funded by rich patrons, complete with magazines, publishing houses, student newspapers, campus organizations, and summer schools where enthusiastic committed cadres were educated and integrated into networks that are still a powerful force in Washington. Meanwhile, liberals, who dominate the universities, remained focused on identity issues and so failed to deliver a real political education or to develop devoted cadres capable of working together in our institutions. Perhaps it is time, then, for Democratic Party patrons committed to a more civic-minded liberalism to follow the conservatives' example and fund independent programs and initiatives to educate a new generation on the left in the principles and realities of democratic politics, and establish a sense of solidarity and common purpose among them.

was perhaps the most radical generation of American citizens since the country's founding. Young people who were incensed by the denial of voting rights *out there*, the Vietnam War *out there*, nuclear proliferation *out there*, capitalism *out there*, colonialism *out there*. The universities of our time instead cultivate students so obsessed with their personal identities and campus pseudo-politics that they have much less interest in, less engagement with, and frankly less knowledge of the great *out there*. Neither Elizabeth Cady Stanton (who studied Greek) nor Martin Luther King Jr. (who studied Christian theology) nor Angela Davis (who studied Western philosophy) received an identity-based education. And it is difficult to imagine them becoming who they became had they been cursed with one. The fervor of their rebellion demonstrated the degree to which their education had developed in them a feeling of democratic solidarity, which is rare in America today.

Whatever you wish to say about the political wanderings of the sixties generation—and I've said a lot—they were, in their own way, patriots. They cared about what happened to their fellow citizens and cared when they felt America's democratic principles had been violated. Even when the fringes of the student movement adopted a wooden, Marxist rhetoric, it always sounded more like Yankee Doodle Dandy than Wagner. The fact that they had taken civics classes taught by high school teachers tapping the blackboard with pointers may have had something to do with it. The fact that they received a relatively nonpartisan education in an environment that encouraged debates over ideas and that developed emotional toughness and intellectual conviction, surely had a great deal to do with it. You can still find such people teaching on campuses, and some are my friends. Most remain well to the left of me but we enjoy disagreeing and respect arguments based on evidence. I still think they are unrealistic; they think I don't see that dreaming is sometimes the most realistic thing one can do. (The older I get the more I think they have a point.) But we shake our heads in unison when we discuss what passes for politics and civic education in our country.

It would not be such a terrible thing to raise another generation of citizens like them. The old model, with a few tweaks, is worth following: passion and commitment, but also knowledge and argument. Curiosity about the world outside your own head and about people unlike yourself. Care for this country and its citizens, all of them, and a willingness to sacrifice for them. And the ambition to imagine a common future for all of us. Any parent or educator who teaches these things is engaged in political work—the work of building citizens. Only when we have citizens can we hope that they will become liberal ones. And only when we have liberal ones can we hope to put the country on a better path. If you want to resist Donald Trump and everything he represents, this is where you must begin.

Acknowledgments

I am blessed with friends from very different political families. I wish to express my gratitude to those, too many to list, who offered comments, criticism, and, yes, solidarity during the writing of this book. A special thanks to the Russell Sage Foundation and its director, Sheldon Danzinger, for graciously hosting me when I unexpectedly took up this project. And to Antonia Blue-Hitchens for crucial research assistance.

This book is dedicated to my wife, Diana Cooper, and my daughter, Sophie Lilla—the Loyal Opposition. And to my old friend Gadi Taub, who years ago urged me to write a book something like this one.

About the Author

MARK LILLA is Professor of the Humanities at Columbia University and a prizewinning essayist for the *New York Review of Books* and other publications worldwide. His books include *The Shipwrecked Mind: On Political Reaction*; *The Stillborn God: Religion, Politics, and the Modern West*; *The Reckless Mind: Intellectuals in Politics*; *G. B. Vico: The Making of an Anti-Modern*; and *The Legacy of Isaiah Berlin* (with Ronald Dworkin and Robert B. Silvers). He lives in Brooklyn, New York.